The Gene Machine

The Gene Machine

How Genetic Technologies
Are Changing the Way We Have Kids—
and the Kids We Have

Bonnie Rochman

Scientific American / Farrar, Straus and Giroux

New York

Scientific American / Farrar, Straus and Giroux
18 West 18th Street, New York 10011

An excerpt from *The Gene Machine* originally appeared,
in slightly different form, in *Scientific American*.

Some of the information in this book is drawn from articles previously published
in *Time*, *The Wall Street Journal*, and *MIT Technology Review*, and on Time.com.

Library of Congress Cataloging-in-Publication Data
Names: Rochman, Bonnie, 1972– author.
Title: The gene machine : how genetic technologies are changing the way we
    have kids—and the kids we have / Bonnie Rochman.
Description: First edition. | New York : Scientific American / Farrar, Straus and
    Giroux, 2017. | Includes bibliographical references and index.
Identifiers: LCCN 2016035006 | ISBN 9780374160784 (hardcover) |
    ISBN 9780374713966 (e-book)
Subjects: MESH: Genetic Engineering—trends | Genetic Testing—methods |
    Genetic Diseases, Inborn—prevention & control
Classification: LCC RB155 | NLM QU 550.5.G47 | DDC 616/.042—dc23
LC record available at https://lccn.loc.gov/2016035006

Designed by Jonathan D. Lippincott

Our books may be purchased in bulk for promotional, educational, or business use.
Please contact your local bookseller or the Macmillan Corporate and Premium
Sales Department at 1-800-221-7945, extension 5442, or by e-mail at
MacmillanSpecialMarkets@macmillan.com.

www.fsgbooks.com • books.scientificamerican.com
www.twitter.com/fsgbooks • www.facebook.com/fsgbooks

1   3   5   7   9   10   8   6   4   2

For Aviv, Shira, and Orli, who couldn't be more perfect for me had they been "designer babies"

The unexamined life is not worth living.
—Socrates

Contents

The Gene Machine

Introduction

He's perfect," pronounced the pediatrician at my two-week-old son's inaugural check-up. I flushed with happiness. What a nice thing for a doctor to say to an insecure first-time mom. And yet he was wrong. We all have genetic mutations. In fact, each human being kicks off life with about sixty new mutations, or changes in their genes.

Some are more obvious than others. Some cause problems, others don't. Nobody's perfect, the adage goes. When it comes to our genes, that's especially true.

In 2007, when I was pregnant with my third child, a routine ultrasound revealed a cyst on my daughter's developing brain. Flat on my back in a darkened room, ultrasound goo smeared on my belly, I froze. Contending with any ominous-sounding "cyst" would have been bad enough, but processing the idea in conjunction with my baby's brain, her tiny body's command center, was chilling. I heard the doctor say that such bubbles frequently go away but may also indicate trisomy 18, a devastating genetic condition that is often fatal soon after birth, if not in utero. I could opt for what doctors call "watchful waiting," keeping tabs on the cyst via ultrasound as my pregnancy progressed, or I could choose an amniocentesis—a needle inserted into my womb—to suction DNA for analysis from my daughter's skin cells, free-floating in my amniotic fluid. In skilled hands, the risk of miscarriage is very low, but it still exists. What did I want to do?

I am an information junkie, so for me the decision was clear: my husband and I went straight from my doctor's office to the office of the specialist who would do the amnio. It was a quick, if sharply painful, assault on my midriff. Predictably, I spent the next day or so panicked that I would miscarry, wondering if I'd made the right choice. Then, on a Saturday afternoon in January, on my middle child's second birthday, the phone rang. The lab was calling to reassure me that initial results showed all the chromosomes where they should be. I sank onto the couch, folded the birthday girl, Shira, in her purple velvet frock and with her budding blond curls, into my arms, and cried with relief.

Imagine my confusion a few days later when our mail carrier shoved a copy of the lab report, which I'd requested, through the mail slot. The amnio had indeed ruled out trisomy 18, but it had inadvertently revealed that my baby had a different, unrelated condition: inversion 9, a transposed ninth chromosome. For some reason, the top portion of the chromosome had landed on the bottom; the segment that was supposed to be on the bottom had migrated to the top. All the relevant information was still there; it's akin to putting your underwear in your socks drawer and your socks in the drawer reserved for undies. You could still find the wayward socks and undergarments, arranged neatly—they just wouldn't be where they ordinarily reside.

It was, I learned, one of the most common genetic errors. Even the lab report dismissed its relevance, its technical language reassuring me that this genetic blooper was benign or, in the language of the lab, "not associated with clinical effects." Perhaps, but I quickly found that knowing about it was strongly associated with emotional effects. One of my daughter's chromosomes was upside down and I wasn't supposed to worry? What's more, it was highly likely that either my husband or I had the same inversion, considering that the lab report noted that this topsy-turvy presentation was considered a "normal familial variant."

To allay my fears, I took the ill-advised yet irresistible path that so many of us tread when faced with unfamiliar health-related information: I paged Dr. Google. I came across one small study that indicated an increased risk of schizophrenia, a finding that still makes me blanch when my daughter, Orli, throws a tantrum or acts particularly irrational, traits that—let's be honest—are not wholly uncharacteristic of the average grade-schooler.

Still, I wasn't prepared for this surprise genetic discovery; nor are countless other parents who encounter similar situations. And yet our growing reliance on a constantly expanding arsenal of genetic tests to enhance our understanding of our children at their most basic cellular level means that this scenario is becoming routine. It's not that I wish I didn't know; I'm glad I know, because now, if new research is published that finds Orli's error correlates with disease, I'll be paying close enough attention to delve into that research and see what I can do, if anything, to reduce her risks. Over the years, I've assimilated the news of Orli's genetic anomaly, incorporating it into the vivid panorama that is my tennis-playing, panda-obsessed youngest child. I can't honestly say that I think about her chromosomal quirk on a daily or even weekly basis, but it does pop into my mind from time to time (remember those tantrums?), and sometimes I wonder: What if the lab report is wrong?

I first started paying attention to the increasing role that genetic testing is playing in children's health back in 2011. I was covering parenting and pediatrics for *Time* and received an email from *Pediatrics*, the official journal of the American Academy of Pediatrics, about an intriguing study. The research explored parents' eagerness to subject their children to tests that transcended the boundaries of the ever-growing catalog of genetic diseases comprising the routine "newborn screening" regimen that takes place in hospitals across the country. The research in question was fascinating: in a large group-practice health plan, 219 parents were offered testing for themselves for genetic variants associated with increased risk of eight fairly common adult-onset conditions—heart disease, high blood pressure, high cholesterol, Type 2 diabetes, osteoporosis, and cancers of the colon, skin, and lung. Asked if they'd like the same genetic testing for their children under age eighteen, the parents indicated that they thought the benefits of testing their kids outweighed any risks. Their gung-ho attitude seemed to stem from a belief that testing could certify their kids as grade-A healthy.

In fact, researchers had tried to talk the parents out of testing their kids (no kids were actually tested; researchers were simply trying to gauge interest) by telling them that there were no known health benefits to doing so, but the parents were not deterred. They may have assumed that their kids—whose average age was ten— would get a clean bill of health. Yet the fifteen different gene mutations associated with the eight diseases the parents were being screened for are so widespread that any particular child would likely have tested positive for nine mutations.

"What? Really?!" I exclaimed to Colleen McBride, senior author of the study and then chief of the Social and Behavioral Research Branch at the National Human Genome Research Institute (NHGRI). Parents, myself included until that moment, seemed pretty clueless about the stories that genes may whisper or shout within our bodies. Yet more and more genetic tests are being introduced—and more and more genes are being singled out as potential troublemakers. We are learning more and more about our genome, our genetic code, but at the same time we still don't understand the basics. As McBride characterized the parents when I interviewed her for the *Time* story: "The more they anticipated feeling good, the more they wanted to test. But the reality is, those parents are going to get bad news. Their kids are going to be at risk for something. So how are they going to react to that?"

That question is at the heart of this book. Is genetic knowledge empowering or fear-inducing, or both? Will it heighten the anxieties of already hyper-anxious helicopter moms and dads, always waiting for the genetic shoe to drop? Will it make parents more diligent about their child's health so that they offer broccoli over brownies and slather on the sunscreen? Will it stress parents out or make them savvier?

There's no indication that parental enthusiasm for genetic information is waning. In fact, a 2014 survey revealed overwhelming support from newly minted parents. In the study, 514 of them were asked within forty-eight hours of their child's birth if they'd be interested in having their baby's genome sequenced, its DNA code deciphered and scanned for errors that may be associated with disease. Thirty-seven percent said they were "somewhat" interested, 28 percent indicated they were "very" interested, and 18 percent reported they were "extremely" interested—a total of 83 percent. Factors such as gender, race, education, and income level didn't seem to make a difference one way or another: four out of five parents thought sequencing their baby's genome seemed like a pretty sound idea.

Even notoriously nonchalant teenagers want a piece of the action. When 282 Cincinnati students in middle school and high school were asked if they'd want to know genetic results about hypothetical conditions that wouldn't affect them for years, 83 percent of them said, "Yes, please."

We are accustomed in our tech-savvy world to accumulating data on our children. There are apps to track a baby's sleep, how often he eats, when she dirties a diaper. But tallying genetic mutations trumps tracking the daily toll of soiled Pampers.

In an information age that compels us to be connected 24/7, no longer do we run to the shelf to pull out a volume of the *Encyclopaedia Britannica*. We are all encyclopedias now with the help of an Internet connection. But all this information can be disconcerting. When I talk about my work, I find that people are either giddy with the promise of information or quaking at the thought of attempting to divine their destiny and that of their children and grandchildren. Even the scientists who are doing the research in the academic trenches, advancing these technologies from theory into practice, are on the fence about what they want to know about themselves and their loved ones. At one of the most esteemed children's hospitals in the country, one geneticist I interviewed spoke proudly of having his genome sequenced; another, whose office is within earshot of the first's, disparaged the trend as self-important and superfluous.

Regardless of differing views, there's no doubt that genetics is reshaping pregnancy and childhood. In the process, it's changing the experience of what it means to be a parent.

◻

Not long ago, having a baby was a fairly straightforward venture. When a couple decided to have a child, they'd ditch the birth control pills and dim the lights. But with no plastic wand with twin purple lines to offer instant at-home confirmation, there was no easy way to gauge success.

The first home pregnancy test—a diagnostic tool now taken for granted by the 4 million women who give birth each year in the United States—wasn't developed until the 1970s. For centuries before that, women relied on various methods, including the so-called "grain test," which supposedly had the added benefit of sussing out gender. An Egyptian papyrus from 1350 B.C.E. explains the not-so-scientific theory behind the practice of a would-be mother urinating on wheat and barley seeds: "If the barley grows, it means a male child. If the wheat grows, it means a female child. If both do not grow, she will not bear at all."

When I was conceived in 1971, my mother learned she was going to become one by heading to her doctor's office after she missed her period. She was offered no genetic tests, got no real-time sonographic glimpse of her daughter-in-waiting doing weightless flips inside her abdomen, had no clue if I'd be a boy or a girl. She didn't know if I'd emerge with ten fingers and the same number of toes, let alone the proper number of chromosomes. She was expecting, and she was oblivious to all the things that could go wrong.

Fast-forward four decades. We are a generation fueled by information, and at no time in our lives do we crave it more than when we are poised to become parents—and once we have our babe in arms. We snap up pregnancy books and memorize developmental milestones as we steep ourselves in the minutiae of a world so unfamiliar it might as well be another country. Even the language is foreign—hyperemesis gravidarum? Reciprocal babbling? As we struggle to find our footing in this strange new place, the cluster of cells nested deep inside is already under scrutiny.

Indeed, at a pregnant woman's first prenatal visit, she rolls up her sleeve and lab techs promptly divest her of multiple vials of blood destined for testing. Soon after, more tests are offered via ultrasound and additional blood draws, and via samples of placental tissue and amniotic fluid, revealing a depth of information that scientists couldn't have fathomed a generation, even a decade, ago. Nor is baby herself exempt.

Across the United States, before parents leave the hospital with their new charges, the infants have taken what a newborn screening education website cleverly dubs Baby's First Test. No one, of course, is suggesting that infants be quizzed on their ABCs. This first test doesn't require much of them, just a couple of drops of blood extracted from a pinprick of their heel. The blood is smeared on specially prepared filter paper, then sent to state labs, where it's screened for rare genetic and metabolic conditions, some of which may quickly prove fatal if not detected.

Newborn screening has been a pillar of public health since 1963, when Massachusetts became the first state to routinely screen infants for phenylketonuria, or PKU, an inherited metabolic disorder that causes brain damage if left untreated. PKU achieved this distinction not because the condition is any more devastating than scores of other genetic diseases, but because it is treatable if detected early—and because the microbiologist Robert Guthrie developed a cheap, effective way to screen for it. This one test spawned the development of others, and what began as a public health measure to detect just PKU has expanded to include dozens of diseases.

There's little question that newborn screening has saved countless lives. Each year, screening identifies more than 5,000 infants who have genetic disorders with severe, frequently deadly, consequences. Babies who screen positive undergo further testing to confirm a diagnosis and are referred for treatment.

The newborn screening program is a government-funded safety net designed to ensure a healthy start for the newest members of society. "Newborn screening occupies a very privileged place," says Robert Nussbaum, former chief of the Division of Medical Genetics at the University of California, San Francisco. "It is a form of testing that undergoes the least amount of scrutiny of any genetic testing we offer in society. Newborn screening is a social contract between the population and the government."

It's a contract of which many parents remain unaware. Although newborn screening is technically optional, parents often have no memory of being asked about it. Detecting sick babies before they have symptoms is so critical for successful outcomes that parents are not specifically asked to consent. Most newborn screening is done in the hospital between twenty-four and forty-eight hours after birth, when a baby's enzyme and metabolic levels are within a measurable, age-dependent range; babies born outside the hospital can have their testing done by a nurse midwife or a pediatrician. "It's an opt-out model," says Natasha Bonhomme, director of Baby's First Test, a federally funded clearinghouse for information and education about newborn screening in all fifty states, administered by the health advocacy organization Genetic Alliance. "It's considered so important that you just go and do it."

That's not the case with the ever-growing array of testing

available prenatally, postnatally, and even pre-conception. The expanding number of options means that the act of parenting, of making choices for and about a child, now starts long before a baby's first breath. Mothers and fathers are urged to leave less and less to chance. More testing, earlier than ever, means they don't have to.

Now, not only do most parents know ahead of time what color to paint the nursery, but technology exists to tell them whether their developing cluster of cells has Down syndrome or a genetic deletion so tiny that it wouldn't have been detected even a few years ago. Mutations that heighten the lifetime likelihood of developing a great variety of diseases can also be identified, a dicey dance considering that many conditions, such as early-onset Alzheimer's, cannot be treated: What good is it to know about a risk of disease if there's nothing that can be done to help? This information exchange about possible or actual progeny is often taking place before the end of the first trimester, maybe even before an embryo implants, forcing parents to make difficult decisions based on an unprecedented deluge of data. Information is usually seen as a good thing, especially in this digital age. But is it possible to have too much information?

◻

Do parents like Laura—flat on her back on an exam table in midtown Manhattan, pants off, shirt on but pushed up over her pregnant belly, silver lamé high-tops glinting in the darkened room—really want to know their child's future?

Laura is thirty-three and has come to see Ronald Wapner for chorionic villus sampling, or CVS, in which Wapner, director of maternal-fetal medicine at Columbia University Medical Center, will plunge a needle through her abdomen into her uterus to collect enough placental tissue (its genetic makeup reflects the fetus's DNA) to determine whether her pregnancy is healthy. He snaps on clear latex gloves, small-talking as he swabs her belly with

Betadine, and compliments her shimmery shoes while warning her she's about to feel a "really strange sensation of pressure." Laura grimaces. Wapner inserts the needle. When he removes it seconds later, the attached syringe is filled with whitish specks floating in a clear pinkish liquid. The white specks are placental cells; they look like goose down. The fluid is tissue culture media that was present in the syringe from the start. "This could not have been easier," Wapner reassures Laura.

No one would make the case that CVS is an enjoyable way to spend a few minutes, but if you're going to do it, wild-haired, charismatic Wapner is the go-to guy. He helped mainstream the first-trimester test and has about as much experience doing the delicate procedure as anyone else in the United States, where it's been offered since the 1980s. It's typically been used to diagnose Down syndrome and other major chromosomal disorders via karyotypes—essentially, chromosome counting—but in the last few years, Wapner has been at the forefront of obstetricians in the United States who are expanding the procedure's purview. In a pivotal study published in 2012 in the *New England Journal of Medicine*, Wapner showed that the tissue collected via CVS can also be scrutinized using a technology called chromosomal microarray analysis. Microarray analyzes fetal cells for countless other, less apparent disorders that occur when a tiny snippet of DNA has been added somewhere it's not supposed to be or deleted entirely, revealing genetic hiccups that previously could not be detected prenatally. Some of these changes are meaningless; others may be associated with autism or rare disorders such as DiGeorge syndrome, which is characterized by heart problems and a roughly 25 percent risk of developing schizophrenia.

This is the latest frontier of reproductive genetics: tests, such as microarray or fetal genome sequencing, so sensitive that they can uncover information no one fully understands. "What should we test a fetus for?" says Wapner. "I'm not suggesting at all that this is bad, but we need to have a discussion about where we're going with this ability. The technology is fantastic. But the easier it is to get information, the more tempted we may be to let our guard down on what we look for."

Microarray is just one of many tests that have been repurposed for pregnancy and beyond in the past few years, as the scope of genetic testing has broadened. In tandem with in vitro fertilization (IVF), more-targeted analysis has allowed women to weed out unhealthy embryos before attempting pregnancy, allowing parents to essentially rewrite their family's medical history: they can stop a genetic disease from climbing up a family tree. Blood-based tests early in pregnancy can now identify Down syndrome and other major chromosomal errors before a woman is visibly pregnant, with unprecedented accuracy, although positive results still require confirmation via amnio or CVS. Genome sequencing—the new gold standard for DNA analysis—lays bare a child's genetic blueprint, including predisposition to disease. (It's particularly useful for sick children for whom doctors have been unable to land on a diagnosis; sequencing can help solve medical mysteries.)

Yet what use could come of telling parents of genetic findings—a gene for elevated Alzheimer's risk found in a six-month-old, for example—that may not be relevant for decades to come? Is it good medicine, not to mention good parenting, to test a first-grader for a mutation that increases the risk of breast cancer, or does it deprive her, as some bioethicists argue, of her right to an "open future"? On the other hand, is it unethical to keep that information from parents, especially if new preventive treatments are developed that would need to be implemented in childhood? Put another way, if you're a parent—or hoping to become one—imagine you're lucky enough to be Angelina Jolie, who in 2013 intensified the national conversation surrounding hereditary susceptibility to breast cancer. The actress announced that she'd had a prophylactic mastectomy after testing positive for a *BRCA* gene mutation, which increases the risk of developing breast and ovarian cancer.

Jolie has three biological children; she must wonder if she's passed on her mutation. If you were her, would you want to know if your daughter has a mutation that sharply increases her risk of breast cancer, years before she even begins to develop breasts?

Society is grappling with a landscape that is constantly remaking itself: tests are replaced so quickly by newer, more comprehensive versions that even professional genetic counselors are finding it hard to keep up. "The type of testing, even from last year to this year, is so different," says Anna Norvez, one of Wapner's genetic counselors, shortly before Laura's arrival for her CVS procedure at Wapner's office. "There is so much that's unfamiliar, but we are all getting more comfortable with dealing with uncertainty. You have to be, if you're going to be offering these tests."

From the room next door where he's dictating charts, Wapner can't help himself. He trumpets: "You deal with uncertainty every freaking day!"

Norvez is unfazed by her boss's outburst. He's prone to passion. Still, he's a good guy to work for; his immersion in research means that Norvez and her colleagues are au courant on the latest developments in genetics. Plus, he's generous: nearly every day, Wapner orders in lunch—Neapolitan pizza the day I was there—and foots the bill.

"I know," she yells back to him, before addressing me again. "There are a lot more databases, so we are increasingly able to give parents some sense of, is this going to be okay or not okay? Obviously, you can't be sure. So it comes down to what the patient's level of comfort is with uncertainty. If they have a very low threshold, they may choose to terminate. If they have a high threshold, they may say, 'I'll try not to think about it, but if the kids have problems, at least I'll know what it might be.'"

In a counseling session before her CVS procedure, Laura tells Norvez that her goal is to set her mind at ease, to be assured she's carrying a healthy baby. A mom of two, she'd lost her third pregnancy at eighteen weeks. "I would like to get this all wrapped up and know the pregnancy is okay," she says.

Norvez nods, then tells Laura that she has an option—microarray—to peer even deeper into her baby's DNA than she'd anticipated. Laura is uneasy. "As a patient, I would assume information like that could be overwhelming," she tells Norvez.

Norvez doesn't try to talk her into it. "That is why someone might not do it," she says. "You've hit the nail on the head. It comes down to your threshold for uncertainty. Would you be okay if we found something unknown?"

"What kind of diseases are we talking about?" asks Laura.

"We can find something on par with Down syndrome, if not worse, which is why some people might say, 'Do it,'" Norvez explains.

"Say I get my initial results and they're good," says Laura. "I'm feeling awesome, I can sleep again, then you call in two or three weeks to say there's a slight deviation of this or that. That's what I worry about."

Like 40 percent of the women Norvez counsels, Laura decides to stick with the narrower spectrum of karyotyping over the more comprehensive microarray testing; her tissue is saved, though, in case she changes her mind.

"It's not bad if you feel this is too much for you," Norvez concludes, then adds her own dry commentary. "We like to overload people with information."

It's true. Having access to more information can enlighten and confuse. It can enable parents to prepare for a child with special needs. Or it can allow them to end the pregnancy before a woman's belly starts to swell. This book is not about right or wrong answers, only extremely personal and intimate calculations. It's not a finger-wagging tale of gloom and doom but an examination of incredible technological advancements. Ultimately, knowledge is power. We are lucky to live in an era when we can dig deep into who we are, although we need to do so carefully and with context from genetic counselors (of which there's a serious shortage). The testing we elect and that which we forgo, and the choices we make based on the results, has profound implications for what sort of society we want to be.

1

How the Jews Beat Tay-Sachs

Screening for Disease Before Pregnancy

When I was pregnant with my son in 2002, I was a gumshoe newspaper reporter in North Carolina covering a group of county commissioner curmudgeons. I earned a living by asking questions. As soon as I'd learned I was pregnant, I'd begun asking about genetic testing for myself and for my husband. A previous pregnancy—my first—had quickly, agonizingly, ended in miscarriage. Before that occurred, I had just begun to cast a wide net, researching which genetic tests I needed. As an Ashkenazi Jew, someone whose family hails from the shtetls of eastern Europe, I was aware of the devastation that can occur when a child is conceived by a carrier couple who both possess the defective gene on chromosome 15 that causes Tay-Sachs, a fatal neurological disease that robs affected children of sight, speech, and movement. Tay-Sachs is an autosomal recessive disease, which means that children conceived by a couple who are both carriers for that disease have a 25 percent chance of being unaffected, a 50 percent chance of being carriers like their parents, and a 25 percent chance of having the disease. Ashkenazi Jews are more likely than others to be carriers—perfectly healthy but harboring the genetic mutation that can prove deadly if paired with a partner's mutation during conception. One of every 27 Ashkenazi Jews in the United States is a carrier, compared to 1 in 250 in the general population. When I met my future husband, he had already been tested as part of an effort at a university Hillel, the organization for Jewish college students. Based on his negative result, my obstetrician said that we were in the clear for Tay-Sachs. But my research had shown there was plenty else to worry about.

In New York City, for example, where awareness of Jewish genetic disease corresponded to the density of the Jewish population, women with my same genealogical profile, descended from the same peasants and farmers and button sellers of Russia and Germany and Poland, were getting tested for an entire panel of diseases more likely to strike Jews who trace their roots to eastern Europe.

Could it really be possible that genetic testing was a function of geography—that a mom in the South, where Jews are a tiny minority, was more at risk of giving birth to a baby with a lethal disease than a mom in Manhattan, where Jews are so much a part of the fabric of the community that a Jewish Genetic Disease Consortium and a Program for Jewish Genetic Health coexist in the same city? I wasn't about to experiment—I wanted testing for the full complement of more than a dozen diseases for which people of my ethnic background were considered at greater risk at the time.

I went on a search for a genetic counselor, a partner in my quest to understand what secrets may lurk in my genes. I found her at a nearby hospital after making multiple inquiries at the state Department of Health and Human Services. She was amused but not surprised when I shared the bureaucratic maze I'd navigated to end up in her corner office. She assured me that I was doing the responsible thing by seeking carrier screening for inherited health conditions—screening my own genes—before I tried to get pregnant again. With the genetic counselor's help, I managed to get insurance coverage for the genetic testing she recommended. (She had advised me to get screened first; if I tested positive for a particular disease, then my husband should also be tested for the same disease.) I had to go to an independent lab to yield up samples of my blood, which was then flown to a lab in Texas for analysis. In 2002, a Jewish girl in the South had to jump through more than a few hoops to peer inside her DNA for markers of risk. My reports came back clear.

More than a decade later, my efforts seem almost byzantine. While some women are still unaware of their risks and options—or know they want genetic testing before or early in pregnancy but still struggle to get access to it—many others are now endeavoring to understand the ever-expanding array of tests made available to them once they're expecting. The first stage of this revolution is carrier screening. Scores of companies and laboratories, recognizing an opportunity, now advocate using newly efficient means of testing to screen all pregnant women in the United States. Many doctors, responding to marketing by these companies, have embraced that approach, offering testing for carrier status for more than 100 serious diseases to all newly pregnant women. Emory University even markets its comprehensive panel as a gift, a present someone can purchase for a beloved pregnant mama. But just because we can screen for more diseases, should we?

There are complicating factors: for one, more-powerful tests can unearth many more gene mutations than traditional means of testing. The mutations, or changes in a gene, might be so unusual that doctors don't have enough information to gauge their severity. As a result, the information can confuse more than clarify.

One company, Gene by Gene Ltd., has introduced a test that screens for more than 250 diseases. "In any genetics market, you'll have people skeptical about that much information," Patrick Miller, then director of clinical and research services for Gene by Gene, said in 2014. "We will be pushing the limits, but we believe in the power of the truly comprehensive test."

Many of the diseases that can be detected are uncommon tongue twisters: aspartylglucosaminuria or pycnodysostosis. Some of them are so rare that doctors may not encounter them even once in a decades-long career, although that renders them no less devastating when they do occur. "If you're the parent and you're caring for a child with a chronic or terminal disease on a daily basis, it's no longer rare for you," says Shivani Nazareth, director of women's health for Counsyl, the San Francisco–based carrier-screening company that is largely responsible for popularizing the concept of "universal" carrier screening, which is pretty much exactly what it sounds like: a vision to screen the universe—i.e., every would-be parent—for the same diseases before or early in pregnancy regardless of where in the world they come from. Of the more than 600,000 people that Counsyl has screened since 2009, at least one-quarter were found to be carriers for a genetic disease. A much smaller percentage of couples that Counsyl has screened in the United States, just over 2 percent, are both identified as carriers for the same disease, which puts them at risk of conceiving an affected child.

⌐

To understand the correlation between genes and disease, it's essential to first grasp what a gene is. Our genome, our genetic code, is composed of DNA, or deoxyribonucleic acid. DNA is the building block of life, but it's probably easier to conceptualize it as a how-to manual, a set of instructions for assembling and operating a living being. A two-stranded molecule that's suspended in the nucleus of nearly every cell (red blood cells don't have nuclei), DNA is our vehicle of heredity.

There are four letters in the DNA alphabet—A, T, C, and G, short for the names of four different molecules: adenine, thymine, cytosine, and guanine. The four molecules, called nucleotides, are strung together in sequences, like cultured pearls on a necklace. It's the order in which they are arranged within an individual's DNA that makes a person unique; this same order, or sequence, directs a body's operations. As with any alphabet, letters combine to make words, which are linked together to form sentences. It's these sentences that form the instructions that we call genes. Genes contain blueprints, or codes, for making proteins. Proteins are the chief operating officers of our cells; they're what make us tick. Different proteins do different things—they carry out functions and give our tissues and organs structure. Some transport oxygen; others sense light streaming into our eyes; still others, called enzymes, calibrate the many vital chemical reactions that keep our bodies running.

Our genes are then bundled onto chromosomes. Mike Bamshad, chief of pediatric genetic medicine at the University of Washington, suggests we think of DNA as an encyclopedia set, with each of our forty-six chromosomes representing a separate volume. But those volumes aren't static, ponderous tomes collecting dust on a shelf: they're dynamic and changing, shape-shifting according to the whims of various inputs from our bodies and our environment. Our bodies grow by making new cells, and each time a cell gets ready to divide into two cells, the DNA in that cell—six feet long, were it to be stretched out end to end—has to make a copy of itself. Invariably, mistakes happen in the process of replication. Amazingly, our cells are able to detect and repair many such errors. Our bodies are models of machine efficiency, continuously problem-solving; reuniting mismatched As, Ts, Cs, and Gs; stitching together frayed DNA strands. But the molecular machinery that detects and repairs mutations isn't perfect, and sometimes, like a sleepy proofreader, it lets a mistake slip by. Usually those changes don't matter much. But sometimes these mistakes, also called mutations or changes or variants, are significant enough that they can cause birth defects or cancer. It's those errors that genetic testing is designed to detect.

Carrier screening focuses on single-gene mutations, which are easily pinpointed because they affect an individual gene. Many of the more than 100 diseases screened for by Counsyl result in birth defects, intellectual disabilities, and shortened lifespans. Some can be treated if detected early. Others have no treatments and are fatal. Several of them are more common in the Jewish community, but even so-called Jewish diseases strike non-Jewish people. One of every 280 babies born worldwide has a genetic disease that could be detected by carrier screening, which makes these recessive diseases collectively more common than Down syndrome.

Counsyl considers itself first and foremost a technology company. Its lab is largely run by robots built by Kyle Lapham, the company's director of lab automation, and a team of engineers. In traditional labs, white-coated techs move from station to station, extracting DNA and pipetting it from one tube to another. They repeat that process for each gene under scrutiny, mirroring the way gene analysis is done in labs the world over. In Counsyl's lab, robots do the bulk of the work. "It is the lab of the future," says Nazareth.

Says Lapham: "When we look to other companies for how we should run, we look to Amazon or Google, not LabCorp." Lapham geeked out showing off his toys when I toured the facilities in early 2016. One, a robotic articulated arm commonly used in auto manufacturing, picks up trays of blood samples to be loaded into a centrifuge, revolves smoothly to the left, and drops its cargo into another networked robot. Many of the pieces that Lapham uses to innovate are 3-D printed in-house. When the articulated arm wasn't speedy enough for Lapham's liking, he and his team built a faster one and filed for a patent. More than 1,000 tubes of blood are processed every day, with each step of the process for every tube recorded on video.

"He can actually watch it from home," says Nazareth.

"Do you do that?" I ask. Lapham shakes his head no.

Despite the emphasis on robotics, people are at the core of Counsyl's work. A team of scientists and genetic counselors mulls which diseases to include among the 100 or so conditions screened for by the company. For the most part, only severe diseases with high detection rates and accompanying treatments or cures pass muster. "At the end of the day, it's about identifying more carrier couples," says Lapham.

Like many companies, Counsyl seeks to drive home the value

of its product—preventing disease—through the personal stories of customers such as Brittany Madore, a hospice nurse. Her experience shepherding desperately ill people through their final weeks of life did not make it easier for her to lose her own son when he was four months old. Her patients' illnesses were rarely preventable; her son's was a different story.

Madore's son, Sullivan, arrived near his due date, an easy, uncomplicated birth after an easy, uncomplicated pregnancy. Sullivan's father delivered him. Their eight-year-old daughter cut the cord.

Sullivan was a great sleeper and an ace breastfeeder. But by the time he was a month old, Madore started to notice that Sullivan wasn't lifting his head as most babies do, nor was he kicking his legs. His cry was meekly quiet, not the four-alarm-fire wail of a typical newborn. By six weeks, when Madore brought Sullivan to the pediatrician, he was "doll-like"—no matter what position she placed him in, that's exactly how he remained.

Concerned, the doctor sent the family to a pediatric neurologist. The night before the visit, Madore went online. She searched and searched for two hours and found that Sullivan's symptoms matched up precisely with those of spinal muscular atrophy (SMA), which causes muscles to wither and, in the worst cases, is fatal in infancy.

Madore described her experience in a blog that Counsyl maintains to share the human faces behind its technology:

> The ride into the neurology office was hell. I knew what we were going to learn at that visit and to have your worst fears confirmed is such a terrible feeling. The neurologist assessed him for about 30 seconds and asked if we had found anything online. I was holding my baby's hand with my face buried on the exam table next to him while trying to enunciate slowly and keep from crying and I said: Spinal. Muscular. Atrophy. Her reply was simple, and I will never forget the reluctance and the sadness in her voice. At the end of a long sigh she softly said "Yeah."

I immediately sobbed all of the tears I had been holding back all morning. I didn't need any other information. I knew she had just given my beautiful 49-day-old son a terminal diagnosis with a prognosis of 4–8 weeks.

Although SMA is the leading genetic killer of infants and toddlers, many people have never heard of it. Madore and her mother, both nurses, hadn't. Neither had Madore's OB or her primary care doctor. And yet one in fifty Americans is an SMA carrier, meaning they have the potential to pass on the disease to their children although they themselves are perfectly healthy. Sullivan was four months old when he died in February 2013. SMA is a death sentence that can be prevented, one whose lethal march through a family can be stopped if it's detected by screening parents for the mutation that causes it.

Every person, you might recall from high school science class, has two copies of every gene, one from mom and one from dad. A recessive disease like SMA occurs when both copies of a gene are altered, or mutated. If only one copy has a change, that person is a carrier, typically with no symptoms.

Mutations can be good, bad, or even inconsequential. Try comparing a mutation to a typo. Depending on where a mutation, a genetic typo, falls within a sentence, it may render it unintelligible. Consider the sentence "I am very intelligent." Leave out every other letter in "intelligent," and the word becomes impossible to read. On the other hand, if you forget to add a second "l," it hardly hinders comprehension; the reader will just gloss over the missing letter and assume the writer spurned spell-check. In a similar way, some DNA errors don't significantly affect the "reading" of the genetic code.

The good news is that many mutations that slip by the repair machinery don't mean much. But some of them turn a key bit of the genetic code to nonsense—garbled instructions that, if carried out, would cause abnormalities in development and functioning.

If such an error isn't fixed and it occurs in a cell destined for the big time—egg or sperm—the genetic change can be passed on to a child, assuming the altered egg or sperm ends up making a baby. These inherited mutations are called *germline mutations* since they come from egg or sperm cells, otherwise known as germ cells. A child with a germline mutation is born with that mutation in every one of his or her cells. With recessive diseases that are detected by carrier screening, the long arm of disease may silently extend generations back. Often, an initial mutation occurred in the egg or sperm of an ancestor and became part of the genetic blueprint of their children, their children's children, and even their future descendants. The mutation didn't present a problem until one of those carriers mated with another carrier.

Other mutations may emerge at some point during a person's lifetime, as a result of environmental exposure or an error during cell division. These are called somatic, or acquired, mutations, and they aren't passed down to future generations. Finally, there's another category, brand-new genetic mutations that are classified as *de novo*, which in Latin means "anew," or occurring spontaneously. De novo mutations may arise during the formation of sperm or egg cells, or in a newly fertilized egg. If such a new mutation occurs in a sperm or egg cell, it may be transmitted to the next generation, beginning a new chain of inheritance. (As men age, the chance of de novo mutations in their sperm cells increases.) If a de novo mutation occurs in a newly fertilized egg, it will be incorporated into every cell of the resulting embryo, including his or her germ cells, and will be passed on to offspring after that person grows up.

Everyone has mutations, maybe even hundreds of them. In fact, in 2012, scientists at the Wellcome Trust Sanger Institute in Cambridge, England, analyzed the genomes of 179 people from the United States, Japan, China, and Nigeria. They discovered that the average person—Gene, let's call him, or Gina—has about 400 errors. The scientists considered most of them harmless, but about

2 per person were categorized as "bona fide" disease mutations, a number that is only expected to rise as sequencing techniques that scan the entirety of a person's DNA—by analyzing the order, or sequence, of the four DNA letters—become increasingly precise.

There are all sorts of triggers and factors that influence development of disease. Family history is one piece of the puzzle, but environment is increasingly understood to play a significant role: smoking or exposure to harmful chemicals, for example, or radiation from the sun. This interplay between heredity and environment illustrates why much of the information contained in your genome isn't black or white. It's indeterminate and gray, a spectrum of murkiness. Sometimes the cause of a disease can be traced to a single gene, but more commonly, as with a disease such as adult-onset diabetes, it can be tied to multiple genes combined with the effects of lifestyle choices. We can't screen a person's genome for these so-called complex diseases—yet. And conditions such as Down syndrome, caused by a third copy of the twenty-first chromosome, are not usually hereditary.

Still, isn't it better to detect whatever we can, especially considering that many of the genetic diseases that can be picked up by testing parents are severe or fatal? Every parent wants a healthy child: we now have the ability to prevent many genetic diseases from being passed on from generation to generation. But despite the arrival of powerful, affordable new technologies and the growing advocacy for their universal use, comprehensive carrier screening is not yet a standard part of the pre-pregnancy/pregnancy routine.

We live in an era that champions prevention. We exercise and fill our plates with veggies because it's good for us. After all, it's easier to take steps to prevent the pounds from piling up in the first place than it is to shed them. Similarly, when it comes to future generations, isn't it more efficient to detect and prevent disease than to cope with the aftereffects, to deal with potential problems on the front end rather than the back end?

Despite the challenges and the snail's pace at which the medical community moves to embrace change, the ability to identify couples at risk of passing on debilitating hereditary diseases (while the risk is still theoretical, before pregnancy shrinks the decision-making landscape to a choice between abortion or giving birth to a child who may be desperately ill) is one of the most obvious ways people can benefit from advances in genetic technology. And yet, carrier screening can be hard to access, as I experienced with my first child in 2002 and as Brittany Madore did in 2013, after Sullivan's death.

Madore wanted to get tested to prepare for future pregnancies, but she couldn't find a doctor in her home state of Maine who was offering expanded carrier screening. She had to travel to New Hampshire, under the impression that she could access testing there, but the doctor didn't believe she needed it and refused to order it. She persisted until she found a nurse practitioner who sympathized; the test results confirmed that Madore and Sullivan's father are both carriers of spinal muscular atrophy—and revealed that he is a carrier for four additional diseases. After Sullivan's death, Madore got pregnant again but had a miscarriage, and she and Sullivan's father split up. "He was really scared of having another baby with SMA," she says. "My opinion is that everyone should be tested. I have no idea what my heritage is. I don't know my father, so I don't have half my history. I could be Jewish. I have no idea."

❒

To imagine what the future of carrier screening might look like, it's helpful to understand its history. Michael Kaback was a young pediatric geneticist at Johns Hopkins in 1970 when, in the course of his clinical duties, he became deeply involved with two families. Harold and Bayla Gershowitz had a toddler son with Tay-Sachs disease. Kaback helped the Gershowitzes find care for Steven. The other couple, Bob and Karen Zeiger, also had a son, Michael, who was less than a year old. Bob Zeiger was a pediatric intern at Hopkins who had done a stint in Kaback's genetics department and had been over to Kaback's house for dinner. Michael appeared to be regressing, and Zeiger asked Kaback to examine him. Kaback had the unenviable task of informing Zeiger that the baby had Tay-Sachs. "It was a devastating disease and an enormously powerful experience as a young doctor," says Kaback, who at seventy-eight is retired from the University of California, San Diego, School of Medicine.

Tay-Sachs is a stealth disease. Newborns develop on a perfectly normal trajectory for the first several months of their lives, doing the yeoman's work of being a baby: neurons firing, neck and core muscles strengthening, eyes exploring the canvas of the world. But by six months, if they could sit up, they no longer can. Their eyes begin to wander and by ten months or so, they suffer convulsions and go downhill day by day. They lose their sight. By eighteen months, they're in an almost chronic vegetative state, bedridden and unable to do much of anything. "It's a horror to watch a child deteriorate like that and not be able to do anything," says Kaback.

When Michael was diagnosed, Karen was well into her second pregnancy. The chairman of Kaback's department thought that the Zeigers should have an amniocentesis to find out whether their fetus had Tay-Sachs, but Kaback disagreed and felt that they should wait until the baby was born. "I was concerned that if Karen found out that she was going to have a second Tay-Sachs baby, she might jump off the roof."

After much deliberation, the couple decided against an amnio. Instead, they decided they would not see their infant until after it was born and had been declared healthy. They couldn't bear to parent two dying children. If their second baby was found to have Tay-Sachs, they would place the newborn in a home care facility or foster care. They would never lay eyes on their child. It was an agonizing decision born of emotional self-preservation, as Kaback recalls, and it presented him with a dilemma. At the time, Tay-Sachs could be detected only in utero or in a baby who had begun to show symptoms. Whether the disease could be diagnosed in an asymptomatic newborn was unknown. It was likely, but had never before been done. Kaback, who had grown close with the Zeigers, knew he had little choice but to do his best to find out.

He began collecting cord blood from healthy newborns to build a bank of control samples so he would be able to compare them to blood samples from babies with Tay-Sachs. In addition, Kaback reached out to John O'Brien at the University of California, San Diego, who in 1969 had discovered the missing enzyme that fuels Tay-Sachs, and Edwin Kolodny at the National Institutes of Health (NIH), who had also helped identify the genetic cause behind the disease. He told them he'd send them control samples in advance of the Zeiger baby's birth; after the baby was born, he'd quickly dispatch to them the newborn's cord blood. Separately but together, the three scientists would try to reach a diagnosis.

The last weeks of Karen's pregnancy raced by. Kaback was in the room when she delivered a baby girl. The moment a child is born is usually one of transcendence, but this time was different. Kaback recalls how Karen and Bob each covered their eyes because they didn't want to know the baby's gender; they wanted to distance themselves until they knew whether their child would share its older brother's destiny. Kaback collected the cord blood sample and raced to his lab. It was 2:00 p.m. He divided the sample into thirds—for O'Brien, for Kolodny, and for himself. By then it was 4:00 p.m. There was a direct flight to San Diego from Friendship Airport (now Baltimore–Washington International), between Kaback's lab in Baltimore and Washington, D.C. This was long before airport security guidelines made it all but impossible for non-passengers to access airport terminals. Kaback sped to the gate for the San Diego flight and gave the Zeiger baby's blood sample, nestled in dry ice, to a flight attendant.

"I said, 'There's a baby's life in flux and there's a handsome doctor who will meet you in San Diego. Will you take it?'" "Of course," she responded.

Kaback called O'Brien to tell him to meet the plane on the opposite coast, then drove directly to Washington to hand-deliver the second sample to Kolodny at the NIH.

The baby's blood held the key to detecting Tay-Sachs, even in a newborn who appeared, by any standards, to be perfectly healthy. Tay-Sachs is caused by a deficiency of hexosaminidase A (Hex-A), an enzyme responsible for degrading fatty-like substances in the nervous system. Without it, the fatty substances aren't broken down, and they build up in nerve cells. This results in the central nervous system disintegrating, which is even more awful in practice than it sounds in theory. Children become completely incapacited and typically die before starting kindergarten.

In 1970, it was not yet possible to sequence a baby's genome. Kaback could only test for the presence of the enzyme. If there is no Hex-A, a baby has Tay-Sachs.

On the day that Karen gave birth, Kaback worked late in his lab at Hopkins, running Hex-A tests over and over. Around midnight, Bob appeared. "Here is Bob, pulling up a chair," says Kaback. "He is like a brother to me by now. He knows his baby and family depend on what in the hell I'm doing."

The clock ticked toward 2:00 a.m. Kaback had run the Hex-A tests eight times, in three different ways, to be certain. Every result had returned high levels of the enzyme. It was a reassuring outcome. Kaback picked up the phone and called O'Brien, who had met the plane around dinnertime in California. Both O'Brien in San Diego and Kolodny at the NIH achieved similar results. "Everyone, using different methods, found loads of Hex-A," says Kaback. The baby, the three researchers concluded, did not have Tay-Sachs.

Bob hugged Kaback. He and Karen still faced the loss of their firstborn; Michael would die before his third birthday. But their new baby, a daughter whom they would name Joanna, would grow up to get her master's in genetic counseling and her Ph.D. in genetic epidemiology, and would place fourth in the triathlon competition at the 2000 Olympics. Together the two physicians walked to the far end of the Johns Hopkins Hospital. First, Bob went to Karen's room in the obstetrics ward and told her the good news. It was 5:30 a.m. As fate would have it, Bob was on rotation as an intern covering the nursery, so when Joanna was born, she had been cared for in the pediatrics ward in a different part of the building. Karen got out of bed and went to the pediatrics ward with Bob, where they held their healthy daughter for the first time.

Kaback left to go home to process the power of what had just unfolded. In the shower, he replayed the chain of events: his relationship with the Zeigers, the excruciating diagnosis he had had to deliver about their son, the heart-thumping efforts to clear their daughter of the same death sentence. It was possible to detect carriers by measuring Hex-A levels in blood. John O'Brien had already demonstrated this. (Carriers have moderate levels of Hex-A; on a scale of 1 to 100, if 100 is a noncarrier and 1 to 5 equates with a diagnosis, 25 to 45 would confer carrier status.) And it was possible to detect affected fetuses via amniocentesis. If carriers could be identified before they conceived, or before a pregnancy was advanced, parents could make informed decisions about having children. They could be in control.

Detecting carriers in the Ashkenazi Jewish population was a daunting prospect. There was no precedent. No one was screening entire populations for genetic diseases. "There was no population-wide carrier screening, period," says Kaback. A framework would have to be created from the ground up. Kaback decided this would be his life's work, to establish population-based carrier screening in the Jewish community. The task would require a massive commitment of money, time, and education. "Most doctors in practice had never heard of Tay-Sachs," says Kaback. "The first thing that Mrs. Rosen is going to do when she hears about screening is pick up the phone and call her doctor and her doctor will say, 'What are these crazy Ivory Tower people saying?'"

In Washington, Harold Gershowitz, who had a child with

Tay-Sachs, called a meeting of friends and colleagues. Kaback showed slides depicting where in the world Jews with the genetic mutation had come from and how it was possible, by identifying carriers of the mutation, to prevent this killer of a disease. He also put up slides of Gershowitz's son Steven's development and steady decline.

"We knew we needed to print up brochures to educate people," says Kaback. The effort would also require supplies, equipment, and personnel. "Harold said he would raise the funds. I showed my slides, I left, and Harold called that night to say he had raised $85,000."

The plan was to enlist the support and infrastructure of the Jewish community to host screenings at synagogues and community centers. Kaback, a secular Jew, secured the support of Baltimore-area rabbis. Press conferences were scheduled, the word went out, and people started calling, eager to sign up. In 1971, the year after Joanna Zeiger was born, 11,000 people were tested. The first screening took place at a Bethesda synagogue on a rainy Sunday afternoon. Fifteen doctors had volunteered to draw blood, and at every station there was a "Band-Aid lady," says Kaback, efficiently bandaging each patient's pinprick wound; 1,800 people showed up to be screened, and Kaback was ecstatic. "It's like you'd written a symphony, never heard it played, [and] then suddenly it was played and it was perfect," says Kaback.

News of the ambitious program spread. Doctors came from Canada, from England, from Mexico, to observe and return to their home countries to set up something similar. Screening programs were established in Australia, in South Africa, in South America, in Europe. Kaback was invited to advise many of these countries, including Israel. A country of Jews had a particular need for a systematic way to screen would-be parents. O'Brien's Hex-A blood test was modified and automated, which made it possible to accurately screen large numbers of people in a short period of time.

Yet opposition came from an unexpected source: the Jewish community itself. A representative of Hadassah, a Jewish women's organization that had been active in helping organize the first screenings in Baltimore, wrote an essay expressing her concern that the Tay-Sachs screening program was stigmatizing Jews, saddling them with the specter of "bad genes." What followed was a meeting at the National Institutes of Health that included the director, Kaback, and members of the Jewish community. Kaback made the case that screening wasn't about eugenics or labeling the population; it was about information. "It wasn't about limiting the population," he says. "It was about having healthy children." The screening program carried on.

What started with the birth of one baby in a hospital in Baltimore has grown over nearly five decades to become a prototype for how grassroots organization, public education, physician buy-in, genetic counseling, and genetic technologies can combine to nearly extinguish a disease in a population. From the time screening began after Joanna Zeiger's birth until 2010, more than 50,000 carriers were identified, among them more than 1,500 couples. As of 2006, when Kaback retired, he had logged 700 Tay-Sachs pregnancies that had been electively terminated due to the pairing of screening with prenatal diagnosis. "Most importantly, more than 2,800 healthy unaffected offspring have been born to these at-risk couples, many of which might never previously have been conceived in such families," Kaback wrote in a journal article about a pilot program to screen for disease within the Persian Jewish community, based on the Tay-Sachs disease-prevention model.

Indeed, the ethical considerations were puny and fairly easily dispatched: few could argue that it was a good thing to bring a child into the world whose brief life would be marked by suffering and inexorable decline toward death.

The success of carrier screening depends upon its uptake. Within the Jewish community, screening programs have reduced the incidence of Tay-Sachs by 95 percent in the United States and

Canada. But there are other communities that have similar carrier rates—French-Canadians and Louisiana Cajuns, for example—in which the importance of screening has not been trumpeted as loudly. People of Irish descent are also at increased risk for Tay-Sachs. Even as a combination of vigilance, dedication, and education has slashed the Tay-Sachs rate among Jews, babies in other communities continue to be born with this terrible disease. Could screening every woman during pregnancy—or, preferably, before—be the solution?

As a disease, Tay-Sachs met critical criteria for the establishment of a major screening campaign: it was serious—so serious that it was untreatable—and detectable within a defined population. A simple, accurate, and relatively inexpensive carrier test existed, making it possible for parents to know their risk of conceiving an affected child. And it was possible to identify and prevent births through the use of prenatal diagnosis and, if desired, abortion. But even Kaback, who was instrumental in making Tay-Sachs the first widely screened-for genetic disease, isn't sure that screening far and wide for multiple diseases makes sense.

As president of the American Society for Human Genetics in 1991, Kaback was resistant to universal carrier screening for cystic fibrosis, which is life-limiting but not typically fatal in childhood. "At that point in time, we had a newish screening test, a major mutation had just been identified, and people were talking about mass screening right away," says Kaback, who favored waiting until the test had proved itself and scientists had further investigated whether the presence of a particular mutation correlated with the severity of symptoms. "Are we going to abort fetuses because they can't run a marathon?" says Kaback. "You get into complicated issues of quality of life." For further reference, he cites Gaucher disease. It's part of an extended panel of tests for Jewish genetic diseases that is routinely recommended to Jewish women in major metropolitan areas, and with apparent good reason: Gaucher disease, type 1 (there are three subtypes) is far more common than Tay-Sachs. It affects about 1 in 450 Ashkenazi Jews, and the carrier rate in this community is a sky-high 1 in 10. Yet the severity of a disease such as Gaucher type 1, which can cause fatigue, anemia, bruising, bleeding, and severe bone pain, doesn't begin to approach the magnitude of Tay-Sachs. "Gaucher doesn't belong in there," says Kaback, who says that its effects can be so mild that there can be "zero symptoms for fifty years." The average age of initial symptoms in the Ashkenazi population—an ache in a hip or a little bruising—is forty-five. Is it right to screen for a disease that isn't fatal?

The Jewish community's response to Tay-Sachs may be a public-health success story, and yet it has led to little standardization in carrier screening, even among Jews. The genetic testing guidelines from the major professional organizations vary considerably from the testing recommended in metropolitan areas with large concentrations of Jews. The American College of Obstetricians and Gynecologists (ACOG) recommends that Ashkenazi Jews be screened for Tay-Sachs disease, Canavan disease, cystic fibrosis, and familial dysautonomia, and the American College of Medical Genetics and Genomics (ACMG) adds five more diseases. But labs that focus on Jewish genetic diseases screen for about nineteen disorders and as many as thirty-eight, with the number continuing to rise.

❑

Peter Kasdan is trying to change that through sheer strength of will. In his first pulpit after he graduated from rabbinical school, Kasdan taught a high school girl with Gaucher disease. "She was always sick," he says. "We always worried about her." As a clergyman, Kasdan knew better than most about comforting the sick. What he didn't know at that time was that having a child with Gaucher—and other diseases that disproportionately affect Jews, leading Kasdan to bury eight of his students over the course of his career—could be avoided. He learned that several years later, in 1975, at the national convention of the Central Conference of American Rabbis, the professional organization for Reform rabbis in North America. During the conference, a resolution was passed to "urge those couples seeking their officiating at marriage ceremonies to undergo screening for Tay-Sachs and other genetic diseases which afflict Jews to a significant degree."

Kasdan was then in his fourth year as spiritual leader of the Reform congregation Temple Emanu-El in Livingston, New Jersey. A self-described type A personality, he took the resolution to heart—and went a giant step further. When couples would ask him to perform their wedding ceremony, he decided he wouldn't just *encourage* them to get tested to see if they were carriers for the genetic diseases that strike Jews more often than the general population. His approach was: "I'm not going to urge them. I'm going to tell them: 'You want me to do your wedding? Get tested.'"

I don't know of other rabbis who refuse to officiate at a marriage if a couple rejects testing, but Kasdan's success rate is impressive. He retired in 2001 as Rabbi Emeritus. Now seventy-five, he has officiated at more than 1,000 weddings. Only two couples declined testing; Kasdan, in turn, declined to marry them. If couples who get tested learn they're both carriers for the same disease, Kasdan, along with a genetic counselor, helps them work through the issues. "The genetic counselor deals with the science of it and I deal with the spirituality of it," he says.

The options, as he sees them: decide not to have kids; choose to adopt; create embryos in a fertility clinic and discard those that have the disease; or get pregnant the regular way, test the fetus, and abort if disease is confirmed. Some couples, he acknowledges, may decide to continue an affected pregnancy. "I have couples who have said to me that it's God's will," he says. "I believe that God has given human beings free will. I would never attribute blame to God if parents knowingly give birth to a child with a fatal Jewish genetic disease. Do they really have the right to give birth to a child whose body will self-destruct? To me, it's a moral issue. But ultimately that will not stop me from standing under the chuppah with them."

Kasdan's maverick, unwavering determination that couples seek genetic testing before they marry (and presumably before they start thinking about having children) has contributed to a significant shift in thinking about the burden of inherited disease. Kasdan hopes to confer the same public awareness that Tay-Sachs has achieved within the Jewish community on the other genetic diseases—including Canavan disease, which killed two of his tiny congregants at ages two and four—that disproportionately affect Ashkenazi Jews. (At least a dozen diseases are more likely to occur in the Sephardic population of Jews who are descended from Spain and from Arab countries, although disease prevalence varies widely by country of origin. Because the U.S. Jewish population skews heavily Ashkenazi, the diseases affecting the Sephardic community have lagged even further behind in the public consciousness.) He is fighting no less powerful an entity than ACOG, the professional organization for the nation's more than 30,000 ob-gyns, which has historically supported carrier screening primarily for diseases whose impact is severe and first observed early in life. While a disease such as Tay-Sachs certainly meets that standard, that's not always the case, as Michael Kaback has noted, for a disease such as Gaucher.

To build his case and educate doctors about the importance of expanding testing, Kasdan, rabbinic advisor to the Jewish Genetic Disease Consortium, leads a medical Grand Rounds program in which geneticists and genetic counselors travel to hospitals to teach obstetricians and pediatricians about hereditary Jewish genetic diseases. He also addresses graduating classes of rabbinical students of all denominations, Reform through Orthodox, advising these freshly minted rabbis to make it a priority to connect with a local genetic counselor wherever they land their first congregation. He delivers his spiel about the importance of catching hereditary disease before it takes its toll on yet another generation and asks if anyone has any questions. Then he lowers the boom: "Now that you know about this," he says, channeling the gravitas of a man who has done for decades the job they're about to embark upon, "if you choose not to advise couples to get tested and then they have a baby with one of these diseases and they choose to sue you, guess who will testify against you as an expert witness?" Kasdan intones this for the shock value, but he says he is fully committed to what he realizes is a "very radical" idea. "If I should ever come across a couple who reaches out to me after they give birth to a baby with one of these diseases and their rabbi did not tell them to get tested, I would suggest they hire an attorney," says Kasdan. "I have no patience for rabbis or doctors who know the reality and don't do anything about it."

While Kasdan is not advocating that a couple who learns they are both carriers scuttle their impending marriage, Dor Yeshorim, an organization that was founded by an Orthodox rabbi who lost several children to Tay-Sachs, is recommending that they think twice. Dor Yeshorim, based in New York City, maintains an anonymous database of genetic testing results. Potential mates—mostly Orthodox Jews—reference the database for what Dor Yeshorim calls a "compatibility check," preferably before "a couple or the parents meet, to avoid unnecessary heartache!" If they're found to both be carriers for the same disease, they are "informed that the match is incompatible." Those couples insistent on pursuing a relationship are offered genetic counseling.

Comprehensive awareness of multiple genetic diseases is slow to trickle down in the States. Not so in Israel, which is a petri dish for Jewish genetic testing. Israel's Ministry of Health continually assesses which new diseases to add to its testing panel; its most recent recommendation lists dozens of diseases according to place of family origin. Jews from Iraq, for example, share some disease risk with Jews from Iran, which overlaps in part with disease risk for Jews from Morocco. In contrast, the United States has no governmental agency devoted to this endeavor. Israel covers prenatal genetic testing for all citizens, including Arab citizens—who are at risk for some of the same diseases as their Jewish neighbors and for some different ones—and strongly advises that testing take place *before* pregnancy. As a result, most couples heed that advice. Just as conscientious young women in the United States begin taking prenatal vitamins in preparation for pregnancy, Israeli women include getting a vial of blood drawn on their to-do list.

Testing before pregnancy can alleviate much heartache and allow for a broader range of options. When Arthur Beaudet, a Baylor College of Medicine geneticist, spoke in 2014 at the widely respected Future of Genomic Medicine conference in La Jolla, California, he told the audience that the time to do carrier screening is before conception. Knowing that many of the inherited diseases result in agonizing deterioration and death in childhood, while others cause debilitating pain, some people would likely change their childbearing plans. The point is that waiting until pregnancy to do carrier screening, if it's even done at all, is too late. "We really need to educate that screening is something you want to do prior to conception," says Beaudet. "From a public-health standpoint, we would like everyone to give birth to healthy children and avoid the birth of children with severe disabilities. If we want to do that to the max, we have to identify inherited risks."

So why hasn't the mainstream U.S. medical infrastructure embraced this framework of pre-pregnancy testing? For one thing, it's expensive for insurance providers to screen every woman of childbearing age, particularly when carrier screening is not yet part of the public consciousness. (The week I visited Counsyl, Nazareth was feeling proud that she'd recently met with Baby-Center, a go-to website for new and expecting mothers, and persuaded the executive editor to include carrier screening on its pre-pregnancy checklist.) For another, many women in the United States—about half, in fact—get pregnant accidentally, which bypasses any opportunity for advance screening. For those whose pregnancies are planned, most make a trip to the doctor's office only after a pregnancy test turns up positive. Even then, many obstetricians—very few of whom have training in genetics—don't emphasize the importance of carrier screening or fully understand which diseases their patients should be screened for.

The advent of sequencing technologies is starting to change whether and how people are screened for genetic disease. From a financial perspective, it's increasingly possible to screen individuals for dozens of diseases for the same price, more or less, as the cost of looking for mutations in a single gene or handful of genes. (And the cost continues to fall: a couple of years ago, Counsyl charged up to $999 for those without insurance; the price is now $349. Many insurance companies cover all or most of Counsyl's cost.) With intermarriage and mixed-race couples increasingly common—broadening the array of diseases that can potentially be passed on to future generations—carrier screening for all seems like a good idea.

After all, in the multicultural world in which we live, ethnicity is no longer the black-or-white paradigm it once was. Bloodlines get muddled into a genetic melting pot in a society in which Asians marry blacks and WASPs marry Jews. "To rely on patients' understanding of their ethnicity and make judgments based on patients' ethnicity is limiting in terms of screening for diseases," says Nancy Rose, a geneticist and maternal-fetal medicine physician at the University of Utah Health Sciences Center and at Intermountain Healthcare, who formerly chaired the Committee on Genetics for ACOG.

☐

That was certainly the case for Sophie-Shifra Gold, whose fourteen-month-old son, Isaac, has Canavan disease, consigning him to a future in which he's destined to deteriorate, both physically and mentally. Gold was not tested before conception or while pregnant to see if she and her partner were carriers for the neurological disorder, usually fatal in childhood, which has left her son unable to support his own head. At an age when most babies are fully engaged in the world around them, babbling and exploring on sturdy legs, Isaac can't sit up on his own.

I met Gold and Isaac at their home in a leafy middle-class Seattle neighborhood on a chilly fall day in 2014. Wendy Marcus—Isaac's grandmother and Gold's mother—signed my questions to Gold, who is deaf, and translated her answers for me. Marcus leads the music program at a local Reform synagogue. Since Isaac's diagnosis, Marcus says she has become "slightly obnoxious. Every day I wear a button that says 'Know Your Genes.'" If Gold, the daughter of a synagogue leader, is not aware of the importance of getting screened for genetic diseases besides Tay-Sachs, the message that Kasdan is trying to promote is clearly not being communicated on a grand scale.

Isaac, propped up in his mother's lap, looked around, his gaze moving from person to person. He smiled often and used his eyes to focus on what he wanted. Gold had baked oatmeal cookies for my visit. When she left the room to fetch the plate and bring it back to the kitchen table, where she, Marcus, and I had been talking, she handed Isaac to me. He wore a red shirt and gray pants with a polar bear on the tush. His head, which was larger than a typical baby's, lay heavy in my arms as I cooed to him. He looked at me with eyes the color of cocoa beans. When he saw that I was a stranger, he began issuing soft cries of protest.

His grandmother spoke of how heart-wrenching Isaac's diagnosis had been for the family. "He will just fade away and start losing his ability to communicate, his expressions, his hearing, his sight. They say it's not painful. It's like an Alzheimer's patient, a slow fade into a twilight zone."

His mother offered a quick smile and urged me not to underestimate Isaac. She said she knew of a few children who had lived into their late teens. Perhaps Isaac would be one of them. She told me that most Canavan kids are wheelchair-bound at some point. Marcus gently corrected her. "Honey, not most—all." Gold nodded resignedly and signed to her mother: "You know that expression,

'A Jew can't live without a miracle'? I will always hope for a miracle." But Gold's hoped-for miracle did not come to pass. A few months after I met Isaac, I learned that he had died.

After Isaac's diagnosis, his extended family got tested to determine who was a carrier. Marcus, Gold's mother, is three-quarters Jewish, while Gold's father is fully Jewish. But it's Marcus who turned out to be the carrier and passed on her carrier status to Gold. Likewise, on Isaac's father's side, the half-Jewish grandmother was confirmed as the carrier. She conveyed her carrier status to her son, Michael Levin. When Levin and Gold conceived Isaac, they each transmitted their gene mutation to their son. As a result, Isaac developed Canavan disease.

Clearly, you don't need to belong to a specific ethnicity to be vulnerable to diseases associated with that ethnicity. Yet ACOG recommends that pregnant women as a group be screened to see if they are carriers for just one disease: cystic fibrosis. Depending upon a woman's ethnicity, testing for additional diseases may be advised.

The buzz surrounding the growing number of laboratories and companies that are marketing expanded carrier screening has prompted a consortium of medical groups representing obstetricians, specialists in maternal-fetal medicine, geneticists, and genetic counselors to address the trend. When, in 2015, ACOG and other medical societies released a statement, their tone was lukewarm, acknowledging only that while "many more conditions, genes and variants are analyzed when expanded carrier screening is used compared with current screening approaches . . . this approach introduces complexities that require special considerations."

The statement is hardly a full-throated endorsement of universal screening. Why the resistance? For one thing, many of the diseases that make up the ever-expanding testing panels have variable phenotypes. A phenotype is composed of a person's observable traits—characteristics such as eye color, but also how a disease impacts a person.

Cystic fibrosis, which clogs the lungs with thick, sticky mucus, is an example of a disease with a variable phenotype. Some people with cystic fibrosis may shuttle in and out of the hospital because the lung disease makes them more prone to illness; those with a milder form of the disease may play on a soccer team with little trouble. There are other conditions that have an even wider symptomatic spectrum—or no symptoms at all. Hemochromatosis, for example, is an iron storage disorder. It's fairly common to be a hemochromatosis carrier, but the disease does not always cause problems even in those who are affected. "Just because you screen positive, you may not necessarily develop a serious condition," says Rose. Under those circumstances, is it always wise to warn parents about possible diseases in their offspring? Won't such warnings contribute to abortions that many might consider needless, or to equally needless anxiety?

In all the excitement over the age of the genome, there's a bit of practical information, a truth of genetics, that often gets overlooked. It's this: DNA is not necessarily destiny. Having a change in a gene associated with disease does not always mean that you will develop the disease. With Canavan, inheriting from both parents a mutation in the gene for a key enzyme equates to a diagnosis. But a woman with a *BRCA1* mutation won't definitely get breast cancer. She has a 65 percent chance of developing the disease by age seventy, but by the same token, she has a 35 percent chance of not developing it.

Another challenge with implementing universal screening is that the rarer the disease being screened for, the harder it is to assess the reliability of the results, because the carrier frequency in the population simply isn't known. As a result, it's difficult to estimate the probability that a result is a false positive or a false negative. "Say we screen you for 100 diseases and you come up as a carrier for a very rare disorder, and we screen your partner and can't identify that in your partner," says Rose. "We don't know if that partner is a true negative because we don't know what the carrier frequency rate is in the population. Adding more diseases to the screening panels means we will find more people who are carriers, but we will not always be able to assess how accurate the results are."

Jennifer Malone Hoskovec, a past president of the National Society of Genetic Counselors, says that pretest counseling is essential, although it rarely occurs. "The more you screen for, the more likely you are to get an abnormal result," says Hoskovec. "You have to decide if this would be meaningful for you or if it would cause increased anxiety. The technology is exciting. We are able to do things now that I never thought were possible even ten years ago. But just because we can do it doesn't mean we need to."

Turning pre-pregnancy carrier screening into the new norm would require nothing short of a transformation in the way pregnant women access care. "I have spent my whole life introducing tests, and doctors don't understand their value and insurance companies won't pay for them," says Beaudet, the Baylor geneticist. Still, he foresees a future, just a generation away, where this transformation will no longer be revolutionary. Classic carrier screening—picking and choosing which diseases to screen for—will simply be rolled into far-more-comprehensive DNA-sequencing tests. By the time my children are having their own children, Beaudet told me, "everyone will do it."

2

Playing God

How Preimplantation Genetic Diagnosis Is Rewriting Family History

I n 2009, Jennifer Davis and her mother, Susan, spoke to a graduate-level bioethics class at Georgetown University. The mother-daughter duo often are asked to do this kind of public speaking, to put a human face on the sorts of agonizing decisions that arise from knowing you're at increased risk of getting cancer. The Davises, who live in the suburbs of Washington, D.C., are part of a generations-long chain of women in their family who carry the *BRCA1* mutation, increasing their risk of getting breast cancer more than fivefold compared to the average woman's 12 percent risk. The same mutation ups their likelihood of getting ovarian cancer to 39 percent in contrast to the 1.3 percent chance of a woman in the general population. The mother and daughter have traced their family's cancer history back to Jennifer's maternal great-grandmother, Ethel, who died of breast cancer in 1930 at age thirty-two. She'd been diagnosed four years earlier and had a crude mastectomy.

The year that Ethel was diagnosed (1926) was three years before the stock market crashed and two before Amelia Earhart would become the first woman to fly solo across the Atlantic. It would be

decades before it was socially acceptable to speak about breast cancer; more than half a century until the geneticist Mary-Claire
King would lead the way to the identification of one individual
gene on chromosome 17 that was wreaking havoc on the breasts
and ovaries of women in families that had witnessed generations
of cancer diagnoses and deaths. Before the discovery, the women
wondered what curse had befallen them. After King's revelation,
they had a name for it—one that King had bequeathed, enlisting
the first two letters from the words "breast" and "cancer": BRCA1.
(BRCA2 was discovered in 1995 on chromosome 13 and, like
BRCA1, was associated most notably with increased risk of breast
and ovarian cancer. More than 1,600 individual mutations, or
changes, have been identified in each of the BRCA genes.)

When Susan gives talks, she brings along a sepia-toned photograph of Ethel, dark hair drawn back into a loose bun, staring past
her husband, in the posed inscrutability that used to characterize
formal photography. Neither is smiling. Susan follows that with a
picture of Ethel's children, Marjorie and Mary Ann, two chubby-
legged, pixie-cut girls in bobby socks and Mary Janes. They were
three and four years old when their mother died.

In 1970, in her early forties, Marjorie, Susan's mother, had a
benign mass removed from one breast. In 1980, Mary Ann, Susan's
aunt, was diagnosed with cancer in one breast. She was fifty-four.
Two years later, she received the same diagnosis in her second
breast. Two years after that, Mary Ann's daughter, Barbara, who was
thirty-eight, repeated the pattern—in 1984, one breast was deemed
cancerous, followed in 1986 by the other. Barbara's cancer spread to
her brain and killed her in 1990, six weeks after her mother, Mary
Ann, succumbed to ovarian cancer. Then, during a surgery intended
to remove her ovaries as a preventive measure, Marjorie was diagnosed with ovarian cancer.

Next, Susan went on to develop breast cancer, and Jennifer
had questionable lumps that led her to choose a prophylactic mastectomy. The bad news came so frequently that it was hard to keep

track of who was getting diagnosed with what, and where, and when. So Susan created a chart. Each time a relative died, she put a black X over the icon representing her. The chart, or *pedigree*, served as pictorial confirmation of what she and her daughter already knew. In short, if you had to pick a family to be a part of based on family health history, this was not one you'd choose to join.

That much was clear to the Georgetown bioethics students, one of whom decided to throw propriety aside and follow up that unassailable line of inquiry: "Say you had the ability to end this legacy," the student said. "Would you?"

As it turns out, that ability actually exists for women considering pregnancy. To save a child from a life clouded by the threat of cancer or the dire certainty of a fatal genetic disease such as Tay-Sachs, parents can turn to a technique called preimplantation genetic diagnosis. With PGD, embryos are created through in vitro fertilization, or IVF, in which sperm and egg are joined in a laboratory. The resulting embryos grow and divide for several days. Then a few cells are removed and tested for the presence of a particular genetic mutation, be it a predisposition to cancer, Tay-Sachs, or a life-limiting condition such as cystic fibrosis. Only those embryos that are free of the mutation are transferred to a woman's uterus (or saved for future pregnancies). The procedure, which can also be used to select for gender, has been deployed for decades to avoid having a child with a serious or fatal disease. More recently, women with identified mutations that heighten their own risk of cancer are beginning to consider it, to spare their future children the same fate.

While genetically vetting embryos to ensure that a child won't have a fatal or debilitating disease is considered morally defensible by most doctors and ethicists, there is of course a slippery slope. It's less clear that using PGD to choose a son over a daughter—or vice versa—for nonmedical reasons is equally justified. The attempt to breed one type of child instead of another, of course, is the basis of eugenics—a kind of misguided quality control for the human population.

It's largely the use of PGD for nonmedical reasons that has sparked concern among ethicists about the prospect of "designer babies." The advent of the gene-editing technology known as CRISPR-Cas9 has further escalated the debate over what degree of tinkering with an embryo is acceptable. (Later in the book, we'll explore these futuristic challenges.)

But even with more straightforward PGD technology, the choice to cherry-pick an embryo is not conundrum-free. The ethics of hand-selecting embryos to eliminate a *BRCA* mutation, for example, are considerably more complex than selecting an embryo free of a deadly Tay-Sachs mutation. While Tay-Sachs always leads to a vastly shortened life span, breast cancer does not necessarily. As we've seen, unlike with Tay-Sachs, merely having a *BRCA* mutation does not automatically confer the disease; it confers increased risk. Plus, there are options—including imaging, surgery, and medication—to detect and treat breast cancer. Still, a *BRCA* mutation means that the chance of developing cancer is undeniably higher.

When faced with the prospect of passing on a faulty gene, more *BRCA* carriers are considering taking advantage of genetic technology to put their family's likelihood of getting breast cancer on par with that of the general population. It's not cheap, though. PGD adds another $6,000 or so to the cost of standard IVF, which can run to $20,000 for just one round. Insurance rarely covers either procedure.

William Schoolcraft of the Colorado Center for Reproductive Medicine, one of the country's premier fertility clinics, has used PGD at least a dozen times to eliminate a *BRCA* mutation in the past three years. "It's by no means 100 percent who are choosing to do it," says Schoolcraft. "I've heard some ladies say, 'My grandma had it, my aunt had it, my mom had it. I want to stop this train.' Others have the attitude that if my baby turns out like me, so be it."

The National Comprehensive Cancer Network observes that *BRCA* status can have a "profound impact on family planning deci-

sions" but doesn't routinely recommend that PGD be discussed with
BRCA carriers. Instead, the guidelines simply note that counseling
may be warranted for those couples who worry about the possibility of passing on a *BRCA* mutation to their children. The guidelines also recognize that PGD requires even couples not struggling
with infertility to use IVF, which is a significant hardship for many.

But support groups such as Facing Our Risk of Cancer Em-
powered, or FORCE, believe it's critical to raise awareness. "We
present it at every conference, we write about it in our newsletter,
and we've done webinars on it, but not everyone knows about
it," says Sue Friedman, executive director of FORCE, whose
members know or have good reason to suspect that they have inherited a dubious family legacy in the form of a *BRCA* mutation.
"It's difficult to make informed decisions about all the choices if
you're not being given all the information."

One study that examined knowledge and attitudes about PGD
found that only a third of women at high risk of hereditary breast
or ovarian cancer had heard of the technique prior to participating
in a related survey. As with abortion, some women think that
PGD should be an option even though they would never pursue
it personally. Others aren't interested because they're past their
childbearing years. "These are very individual decisions," says
Friedman.

That day at Georgetown, Jennifer Davis pondered the question
the student had posed about whether she'd select an embryo that
did not have the mutation she was born with. It was a very personal
question. But as an outreach coordinator for FORCE, Jennifer
was used to fielding nosy queries. The curiosity didn't faze her.

She decided to address her mother, not the student, in response.
"Basically," Jennifer observed, "if you had the option to preselect
embryos to not have a mutation, you wouldn't have me!"

"I wouldn't do that!" Susan replied, slamming her hand down
on a table. The students were riveted as this real-life drama played
out in their classroom.

But Jennifer, who was twenty-three at the time and had no children yet, told the class that she would.

Susan's knee-jerk negative reaction was understandable. It was incomprehensible for her to imagine that had she chosen not to have a child with a *BRCA* mutation, Jennifer—the daughter she has raised and loved for three decades, her only child now that her son, Richard, had died in a motorcycle accident—wouldn't be here. (Richard had also inherited the gene mutation, and had become one of the first men to speak publicly about it.) But for Jennifer, the choice would be far less agonizing, akin to the same choices that any woman going through IVF makes. It would be a means of protecting her future children from experiencing the losses to cancer that her family had endured.

Fertility clinics "grade" embryos; the heartiest ones score the highest. PGD adds another criterion for selection: only those embryos without a mutation would be potentially available. "My mom didn't have that choice," Jennifer told the class. "This disease and this mutation robbed me of a life with my grandmother. She died when I was twelve. And my mother grew up without a grandmother as well. Legacy is important to me. If I can potentially eliminate it running through my family, I would prefer to do that."

For someone like Jennifer who knows she has a *BRCA* mutation, conceiving naturally would mean that she'd have a one-in-two chance of passing the mutation on to her child (a father with a *BRCA* mutation has the same 50 percent chance of transmitting it to future generations).

❐

Those odds were unacceptable to Deena Kobell. In 2005, the Philadelphia attorney wrote to doctors in Belgium, England, New York, and Israel asking if they could help her conceive a child who didn't have the same *BRCA1* mutation that had dogged generations of her family. Kobell, who was diagnosed with breast cancer in 2001 at age twenty-nine and treated successfully, was determined to ensure that her children wouldn't face the same fate. "I got letters saying nope, it's not possible, you can't do it," she told me.

On the verge of giving up, she connected with Mark Hughes, founder of Genesis Genetics, pioneers in the field of PGD, in a suburb of Ann Arbor, Michigan. Hughes had partnered in the first-ever use of PGD for a specific disease, cystic fibrosis, in 1991. He'd never performed detailed analysis to choose an embryo without a breast cancer mutation, but when he called Kobell at home, he said he was willing to try. Kobell was thrilled, despite the tension it subsequently caused among family members who share the mutation but worried that what she was doing was akin to playing God. For his part, Hughes reserved judgment. "If you have a family that has lost all sorts of folks to breast cancer, it's quite evident that the mutation is pretty serious for your family," he told me. "Is it appropriate to test embryos for a mutation that may never actually cause disease in that individual? If a family asks to avoid it, that's what we do." The technology worked: Kobell's daughter, Eve Helena—named for Kobell's mother, Helen Evelyn—is proof. Believed to be the first child in the world deliberately created free of a breast cancer mutation, she was in second grade when I met her at her family's restored four-story row house in Philadelphia.

Eve perched on a kitchen stool as Deena recounted her story. "Remember I told you that when you were born we made sure you weren't going to get breast cancer?" Deena asked Eve. Eve looked interested.

Deena turned toward me. "It came up in the context of my mother. Eve was asking what my mom died of. She had ovarian cancer and Eve asked if I would get it too. I said no because I had my ovaries removed. She knows that I've had breast cancer because she's seen my scars. And I said, 'When you were an itty-bitty tiny little thing, we tested to make sure you didn't have the gene that Grandma Helen and I had.' We made sure you won't get breast or ovarian cancer.'" (Deena's simplified explanation reflected her daughter's young age, but it's not entirely accurate. While PGD ensured that Eve did not inherit a mutation that ratchets up the risk of breast cancer, it can't eliminate the possibility that she could one day be diagnosed. Eve's risk is on par with that of any female in the United States, one in eight of whom develop nonhereditary breast cancer at some point in their lives.)

Eve piped up. "What happens when you get cancer?"

"Well, you get sick," said Deena. "You could be in serious trouble."

"You mean die?"

"Yeah."

Eve, the dark-haired, dark-eyed great-granddaughter of Holo-caust survivors, began life as one of a group of fifteen fertilized embryos. By the time the embryos had grown for four days—long enough for Hughes to biopsy them to look for the presence of *BRCA1*—five had stopped dividing, leaving ten for analysis. Eight of the ten were able to be biopsied, revealing that just four didn't have the mutation. By the embryos' fifth day in the lab, only three remained viable. All three were transferred into Deena's uterus. Nine months later, just one made it out: Eve.

"Now I'm relieved," Deena told me. "I have a lot of things to worry about, but I can cross that off the list."

Eve, who takes piano lessons and dances ballet on Saturdays, had listened quietly to her mother recount how she came to be. "I'm not quite sure what a seven-year-old can understand," Deena said.

Eve's favorite subject is math. She loves neon colors. Her favor-ite spot in her pink room with butterflies on the wall is her bed, which has a lavender mermaid comforter and an Emily Windsnap book that Deena had just bought her. Eve and I bonded over the book, part of a series that revolves around mermaids in general and Emily, who is half girl, half mermaid, in particular. I know this trivia because across the country in Seattle, where I live, I am reading this same series with my own second-grade daughter.

Eve's birth—her very existence—came about because of the availability of a test for the *BRCA* mutation. Does this make her a designer baby? Despite her unusual origins, it struck me that she's no different from seven-year-old girls in general—my daughter included.

Not everyone can afford to choose PGD, nor does everyone want to. In response to an article I wrote for *The Wall Street Journal* in 2014 about using PGD to sidestep the possibility of transmitting a breast cancer mutation, readers' comments were sharply divided on the propriety of deciding that because someone *may* get a disease down the road, she or he shouldn't exist. Since Eve's birth, Genesis Genetics has helped more than 380 couples have children without a *BRCA* mutation. There are no federal rules governing PGD, putting individual scientists and laboratories in charge of deciding where—or whether—to draw the line.

Not surprisingly, parents who pass on the mutation often feel tremendous guilt. "Though parents know that children are hostages to fortune, how can mothers bear to realize that they have unintentionally tendered their daughters a curse, a ticking time bomb, or a seed of death?" writes Susan Gubar in *Memoir of a Debulked Woman*, a book about Gubar's experience living with ovarian cancer. Fathers, of course, can just as easily pass on a mutation. For them, the guilt may be even worse. Say that the mutation has been transmitted from great-grandfather to grandfather to father and then on to that father's daughter. Since men with a breast cancer mutation have a much lower, albeit increased, risk of developing the disease, the presence of the mutation may come as a complete surprise. This happened recently to one of my best friends.

As the bioethicist Arthur Caplan has noted, genetic information is "exquisitely sensitive." Worrisome genetic test results can make people feel that "they harmed their children, or that they themselves are flawed. Most people don't feel that way if their kidneys start working inefficiently. They don't say, 'I'm a flawed human being.' But they feel that way about their genes."

Gwendolyn Quinn, a psychologist at Florida's Moffitt Cancer Center who led the study that assessed attitudes about PGD, says the cost and the hassle of going through IVF are two reasons why the option isn't embraced by many *BRCA* carriers. Those same factors, says Quinn, make it unlikely that PGD will be widely adopted to select mutation-free embryos, let alone to choose embryos with desirable traits. "Some people worry this will lead to designer babies, but to me that is a leap," she says.

Quinn began researching viewpoints about the use of PGD for hereditary cancers in 2008, starting with examining health care providers' perspectives. She found that oncologists didn't agree on whether embryos should be discarded because they have a mutation that increases their risk of one day developing cancer.

A year later, Quinn conducted a large survey targeting all members of the FORCE community. Of the more than 900 women who responded, most were over age fifty and had completed their families. Still, a third said they'd want to consider PGD if they were planning to conceive, whereas 38 percent said they wouldn't and 29 percent selected "don't know." Judging from the respondents' comments, many shared Susan Davis's outrage at the existence of the technology. "They said, 'If this technology were available when my mom was pregnant with me, I wouldn't be here.' They said, 'This is not a death sentence. I had breast cancer and I survived.'"

But, like Jennifer Davis, the small subset of women who had not yet had children seemed open to, even excited about, the possibility of going through IVF in order to have a child they knew would not carry this same mutation that had afflicted generations. That enthusiasm carried over to a focus group of thirteen women in their twenties that Quinn assembled in 2010. The women had a *BRCA* mutation but had not been diagnosed with cancer. Half of them had undergone surgery to remove their breasts and/or ovaries, and the other half said that they were considering it. The majority of the women said that they'd had no idea that PGD existed. "They were angry that no one had talked about it with them as they were making treatment planning decisions," says Quinn.

For those women in the group who had wanted to become

mothers, learning about PGD was a revelation. "Many of these women had said: 'This cancer is going to end with me. I'm not going to have children,'" says Quinn. Yet when they heard about PGD, they were no longer as certain. "There were great quotes from this focus group in one of my papers from a woman who said, 'I always dreamed I would get pregnant the old-fashioned way, sitting by the fire, drinking a glass of wine. PGD sounds so clinical.'" But to reduce her cancer risk, the woman was going to move forward with having her ovaries removed, so she would need to undergo IVF to get pregnant regardless. Why not do PGD also?

Genesis Genetics, which was responsible for Eve's birth, performed 762 cases of PGD in 2015, scanning embryos for more than 150 different genetic disorders. The workload has skyrocketed in recent years due to increased awareness of genetic testing for mutations; as recently as 2012, Hughes did just 128 cases. PGD, which tests embryos for only one specific inherited condition known to affect a family at a time, is a small subset of preimplantation genetic testing; at Genesis Genetics, 90 percent of its preimplantation genetic testing is focused on chromosome analysis because the right number of chromosomes dramatically improves the odds of a successful pregnancy. "With PGD, we test for only the disorder that they know about," says Hughes. "We don't screen for anything that might show up. That's mostly because if we did such screening for any and every thing, we'd not find any 'normal' embryos. We all have bad genes and are blissfully unaware. The goal here is to avoid the disease that we know about."

❐

PGD has its roots in the 1930s, when attempts were made to select for the gender of agricultural animals. In the late 1960s, scientists— including Robert Edwards, who would go on to win a Nobel Prize for developing IVF—succeeded in identifying gender in a rabbit blastocyst, a ball of cells just a few days old. What was possible with bunnies took considerably longer to translate to babies.

In order to use PGD therapeutically to eradicate the possibility of disease, IVF first had to emerge onto the reproductive scene. It did, in the late 1970s and early 1980s. In 1990, the first use of PGD in humans took place at Hammersmith Hospital in London when Alan Handyside and Robert Winston successfully identified the gender of embryos from couples who wanted to avoid passing on an X-linked disorder such as hemophilia. These disorders typically affect boys, who, unlike girls, don't have a healthy second X chromosome to compensate.

Shortly after, Mark Hughes, the doctor who made Eve's existence possible, teamed up with the Hammersmith group to perform sophisticated DNA mutation testing for inherited disease, which opened the door to testing for gene mutations on all the chromosomes. In 1991, they performed PGD for cystic fibrosis, which is caused by an error in one gene. Cystic fibrosis had been identified just two years earlier on chromosome 7. "This was a common and terribly dreadful disorder that made sense to target as the first PGD disorder," says Hughes.

Now, there is virtually nothing in the human genome that can't be tested for, as far as disease is concerned. "The technology represents the limit of medical diagnostic testing forever because you're testing one cell, which is the smallest unit of life, for one gene, which is the smallest unit of inheritance, for one typographical DNA mutation in 6 billion letters of DNA," says Hughes. "And you can have the answer tomorrow morning." PGD has been performed for more than 700 mostly lethal inherited disorders. "The question is no longer, can we test for it? The important question is, should we? There is a distinct difference between a genetic disease like cystic fibrosis or muscular dystrophy and a trait like hair color or dry/wet earwax."

A single letter in the genome, for example, results in "dry," gray wax in Asians or cola-colored wax in Caucasians. "It's fun, but it isn't the kind of thing that someone would test for," says Hughes. "This is genetic entertainment, not genetic medicine."

Even attitudes toward PGD for cystic fibrosis have evolved as better treatments have increased longevity. Fifty years ago, the sticky mucus built up in lungs, regularly killing affected children long before they reached double digits. With modern pulmonary medicine, the disease is not as horrid as it used to be. But it's still bad. It's still life-limiting. (Of course, it can be argued that the very experience of being human is life-limiting.) Should people with cystic fibrosis be grateful that they get to live "into their 30s, 40s and beyond," as the American Lung Association puts it? As they approach those ages, their thoughts may turn to childbearing. Should they choose PGD?

Genetically selected embryos raise the specter of designer babies. But what is a designer baby, anyway? The phrase "designer baby" is often bandied about pejoratively to refer to "enhanced" embryos from a dystopian future, ones that are created to have certain traits—blue eyes, perhaps, or blond hair. Although those traits are not currently a mainstay on the menu at fertility clinics, a few centers claim to offer this service. More commonly, clinics are increasingly allowing parents to select gender for any reason at all. Personal preference is enough. The United States, in fact, is a hub for "reproductive tourism." Along with Mexico, it is one of only a few countries that allow gender selection for nonmedical reasons, as well as to eliminate genetic disease. In 2006, an online survey of U.S. fertility clinics that offer PGD revealed that 42 percent had performed sex selection for nonmedical reasons, and almost half of those placed no restriction on the circumstances (such as limiting it to a second or subsequent child). In 2015, the American Society for Reproductive Medicine formalized its position, saying that fertility doctors have "no ethical obligation to provide or refuse to provide nonmedically indicated methods of sex selection." In other words, the decision is left up to individual clinics. Meanwhile, the American College of Obstetricians and Gynecologists takes a more absolutist view, stating that it opposes the use of PGD to select embryos for social or cultural reasons.

Identifying gender in an embryo is easy. What's more challenging for geneticists is figuring out what predisposes someone to have perfect pitch or a facility for calculus. Numerous genes influence the expression of characteristics and traits such as height, high IQ, perfect pitch, or coordination. Environment, hard work, and luck play roles too. We can't as yet infuse an embryo with intelligence or musical ability or a killer golf swing. Nor should we, says Hughes. "Just because we can do something doesn't mean we should."

That hasn't stopped society from fretting about the possibility. With each advance in gene technology, the hand-wringing intensifies, as do worries that designing babies will create a world of haves and have-nots. That worry is nothing new. In 1998, *The New York Times* printed a letter to the editor expressing concern that research into the connection between genes and intelligence invited abuse by those parents "intent on having their designer baby be more intelligent than the children of others . . . Will all newborns be in vitro someday so we can screen out the embryos that are . . . likely to produce children with I.Q.'s below 160?"

Many years have passed since that letter appeared in the paper of record, yet we still don't know how to build a Baby Einstein. In this sense, at least, the fuss over designer babies may be overstated. "We don't have the genes nor do we know how to select for them to create designer babies," says Jamie Grifo, director of the Division of Reproductive Endocrinology and Infertility at NYU Langone Fertility Center.

Grifo was among the first doctors to use PGD in the United States, to help a family avoid hemophilia in their children. "The ethics of the way we talk about it has changed a lot," says Grifo. "It has become about personal choice."

Given the choice, do you want your child to have the same problem that you do? Deena Kobell, for one, did a Google search on her name and found that an academic had written a paper about her decision to use PGD. "People commented, 'Oh, designer babies,'" she says. "But that's not what I did." What she did, she says, is give herself peace of mind. "When you find out that you are *BRCA+* and you can still have children and not pass that on, it's much less of a blow."

"You have to think about it from the perspective of the woman whose life has been thrown upside down and the guilt associated with choosing to have a baby that could have those same problems," says Grifo. "There are ethicists and people who judge, but they don't really get to vote. I always say, Practice empathy first and then judge later. If you imagine it's you, all of a sudden it's not so simple."

Depending upon how you define designer babies, it can be argued that they already exist within the world of fertility treatment. People who rely on egg or sperm donors to have a child usually don't just get the luck of the draw. They intentionally select a donor based on specific criteria: looks, personality, and level of education, among other options in the drop-down boxes in many online forms.

Famously, there was a Southern California sperm bank, the Repository for Germinal Choice, that solicited sperm from Nobel laureates and garnered headlines when it opened in 1980. Nicknamed the "Nobel Prize sperm bank," it was the place to go if you sought super-sperm. Yet two of the three laureates who agreed to sign up ultimately balked, and the third—William Shockley, who won the Nobel Prize for physics in 1956—donated just once. The bank then decided to pivot: "Instead of recruiting Nobelists, I decided to predict who the future Nobel laureates would be," Paul Smith, the bank's first director, told the journalist David Plotz about how he and Robert Graham, the repository's founder, went about soliciting donors. As Plotz wrote:

> He approached young scientists who had won awards. He haunted the campuses of University of California at Berkeley and Caltech, where young *Über*nerds are thick on the ground. At first, Smith and Graham focused on hard scientists and cared only about intelligence, but they soon

realized their clients weren't satisfied with just brains. "Women would always ask how good-looking he was and how tall he was, and they would want to know if he was athletic. We realized that if you are going to offer choice, you have to offer women a *real* choice," Smith says.

. . . Repository literature did brag incessantly about the A-one sperm, but most clients seem to have recognized that this was not exact science. They hoped for a slight boost, not a mini-Nobelist.

These women were onto something. Rather than picking and choosing particular genes to try to make a child smarter, more beautiful, and a terror on the basketball court, the best way we know to have a smart, good-looking, athletic child is to procreate with someone who is smart, good-looking, and athletic. Indeed, this was the mothers' line of thinking, according to conversations Plotz had with nine sperm bank recipients.

If you're an Uma Thurman or Ethan Hawke fan, you may recall seeing the sci-fi flick *Gattaca*. That film portrayed a world in which "perfect" children are conceived with the aid of genetic engineering. People conceived the old-fashioned way are destined for menial jobs; they are dubbed "in-valids" or "de-gene-rates." Essentially, everyone who is anyone is a "designer baby" in the film. In his book *Redesigning Humans*, which was published five years after Hawke connived against his "inferior" genetic makeup in the movie, and prevailed, the biotech entrepreneur Gregory Stock revives *Gattaca*'s argument: in the future, he suggests, reproductive technology will be the only true option for giving rise to the next generation. "With a little marketing by IVF clinics," he writes, "traditional reproduction may begin to seem antiquated, if not downright irresponsible. One day, people may view sex as essentially recreational, and conception as something best done in the laboratory."

Yet from his perch performing hundreds of PGD cases a year to prevent genetic disease, Mark Hughes rejects that sort of future as impractical. His company, Genesis Genetics, has two labs in England in addition to ten others in countries including Jordan, South Africa, and Brazil. For every case of PGD in England, the company was required by the United Kingdom's Human Fertilisation and Embryology Authority (HFEA) to submit what he described as a "white paper" detailing the severity of the disease, the impact on the particular family, the available treatments, and the exact technology they planned to use in order to conceive an unaffected child. The HFEA, a government body that monitors U.K. fertility clinics and research involving human embryos, would then issue a ruling in favor of or against the use of PGD on a case-by-case basis. "The agency finally realized that they were approving every case," says Hughes. "No one was asking to go through all this for trivial reasons. No one in their right mind would go through the emotional roller-coaster, hoop-jumping of IVF for something they didn't need to do. Twenty-five years of PGD have shown that this is too complicated for people to turn to for designer baby crap."

What constitutes a designer baby is a "squishy" concept, says Misha Angrist, a Duke University researcher who was among the first people to have his genome sequenced and made public, back in 2009, as part of Harvard Medical School's Personal Genome Project. (The project, which is ongoing, collects and shares genome data as a means of accelerating research and because "sharing this data is good for science and good for society.") If you have more than forty repeats of a certain gene sequence, you will develop Huntington's disease, which is an agonizing, personality-changing way to die. Is an embryo selected for its absence of those deadly repeats to be considered a designer baby? Perhaps technically yes, but not in the more fanciful way that some couples choose XX embryos because they dream of selecting a ruffled layette. "It's a loaded term, a cliché, shorthand for *Gattaca*," says Angrist. "It has a lot of baggage. When my students say, 'I want to write a paper about designer babies,' my first impulse is to say, 'No, please don't.'"

Is a designer baby one that is created to possess certain characteristics, or, perhaps, not to possess them? I'd argue that we're currently more equipped, through PGD, to eliminate traits than to inculcate them. What qualifies as undesirable is, of course, subjective. And sometimes parents' choices can be especially surprising.

In an intriguing spin on the designer baby conundrum, some couples intentionally select an embryo with what many people would consider a deficit. Deaf couples have requested to do PGD for the express purpose of having a deaf child; dwarves have sought to have children who are "small people" just like them. In 2002, news broke of a deaf lesbian couple from Bethesda, Maryland, who enlisted as sperm donor a profoundly deaf family friend with five generations of deafness in his family. Candace McCullough and Sharon Duchesneau first sought assistance from a sperm bank but were informed that congenital deafness disqualifies a potential donor.

The women wanted to have children who would share in deaf culture, and they had two: a daughter, who is profoundly deaf, and a son, who is less so. While they received some support for engaging a deaf sperm donor, they also fielded ample criticism. "I can't understand why anybody would want to bring a disabled child into the world," said Nancy Rarus of the National Association of the Deaf. She added that deaf people "don't have as many choices."

In 2008, in an article in the journal *Fertility and Sterility*, the same survey that tallied the number of U.S. fertility clinics that provide nonmedical sex selection also reported that 3 percent of PGD clinics surveyed said they had used the technology to allow clients to choose an embryo with a disability. The vast majority of fertility doctors don't sanction PGD for that purpose. "In general, one of the prime dictates of parenting is to make a better world for our children," Robert Stillman of Shady Grove Fertility Center in Rockville, Maryland, told *The New York Times*. "Dwarfism and deafness are not the norm."

Indeed, PGD to prevent dwarfism can be a lifesaving technology. Dwarfism rates vary, but the most common type, achondroplasia, occurs in 1 of every 15,000 to 40,000 births and affects growth, particularly in the long bones of the body. People of average height can have dwarf children, just as dwarves can give birth to children who grow to tower above them: when both parents in a couple have dwarfism, their children have a 50 percent chance of being dwarves by inheriting one dwarf gene from one parent and one unaffected gene from the other parent. They have a 25 percent chance of reaching regular height should they receive one unaffected gene from each parent. And they have a 25 percent chance of inheriting a fatal "double-dominant" mutation in the unfortunate event that they wind up with two dwarf genes—one from each parent. Using PGD can identify an embryo with the double-dominant mutation. Such babies won't survive, and PGD can clearly save the parents much heartache.

In 2003 in France, Stéphane Viville conducted what appears to be the first documented case of PGD for dwarfism. He used the technology more conventionally, to select an embryo without dwarfism, in the case of a couple where one person was a dwarf and the other not.

"Interestingly," writes the pediatric cardiologist Darshak Sanghavi, "if confronted with a situation where both parents were dwarves, Dr. Viville says that he most likely would implant only an embryo destined for normal height—and forbid not only double dominants but also dwarf embryos."

Sanghavi takes an entirely different view. "I think Dr. Viville fears that PGD could be used willy-nilly to make genetic freaks," he writes:

> Yet the same fears pervaded the issue of in vitro fertilization decades ago. The small number of PGD centers selecting for mutations doesn't bother me greatly. After all, even natural reproduction is an error-prone process, since almost 1 percent of all pregnancies are complicated by birth defects—often by more disabling conditions than dwarfism or deafness.

More important, as a physician who helps women dealing with complex fetal diseases, I've learned to respect a family's judgment. Many parents share a touching faith that having children similar to them will strengthen family and social bonds.

❐

People with all sorts of disabilities (whether or not you consider dwarfism a disability) lobbied for years for legally enshrined accommodations under the Americans with Disabilities Act. These accommodations aren't meant to set them apart; they are intended to level the playing field so that people with disabilities have the same rights and opportunities as people without them.

"Having fought so hard and so long for the accommodations and respect and understanding they need to become full and equal participants in society, people with disabilities now face the prospect of elimination at the hands of people who don't understand that their lives aren't diminished—just different," writes Dan Kennedy in *Little People: Learning to See the World Through My Daughter's Eyes*. The daughter he references is his oldest child, Rebecca, who was diagnosed with achondroplasia a week after she was born in 1992. This passage appears in a chapter of his book with the distressing title "The New Eugenics."

To understand why Kennedy called his chapter "The New Eugenics," and why its specter disturbs so many, it's first necessary to understand what the "old eugenics" were.

Historically, eugenics—*eu* from the Greek for "good" or "normal" and *genos*, meaning "birth"—refers to the effort to build a better race, to jettison genetic diversity in favor of a perceived superior human prototype.

Eugenics, of course, hit its nadir with Hitler's campaign to exterminate the Jews. In addition to the systematic gassings they inflicted at their network of concentration camps, the Nazis also sterilized more than 400,000 people. As they worked to exterminate not only Jews but Gypsies, homosexuals, and the disabled, they escalated the ideals of genetic purity to a horrific and unprecedented level. But the germ of the idea that certain people are superior and thus should make babies together has roots stretching back to Plato, who proclaimed in *The Republic* (attributing the words to Socrates): "The best men must have intercourse with the best women as frequently as possible, and the opposite is true of the very inferior."

Eugenics got its start, at least in name, in 1883, when Sir Francis Galton, a British polymath and cousin of Charles Darwin, first used the term to refer to the propagation of the "well-born." His focus on giving "more suitable races or strains of blood a better chance of prevailing speedily over the less suitable" was endowed with scientific context by the laws of heredity that had been articulated by Gregor Mendel, who conducted genetic experiments with pea plants. For God-fearing folk, there was even so-called divine evidence in favor of culling unsuitables from the human race in the Bible itself. When Moses bestows the Ten Commandments upon the Israelites, he cites this divine pronouncement: "For I, the Lord thy God, am a jealous God, visiting the iniquity of the fathers upon the children unto the third and fourth generation of them that hate me" (Exod. 20:5). Iniquity, whatever that meant, appeared to warrant a Biblical curse, devastatingly revisited over and over again on future generations. The perceived connection between heredity and all sorts of social, intellectual, and moral shortcomings defined the field of eugenics as it took root in America.

Beginning in the early 1900s, things started to get competitive, with "Better Baby Contests" held at state fairs to separate the infant wheat from the chaff. Tykes were scored using assessments that considered physical features alongside intelligence. Pint-size contestants offered up heads, chests, arms, and legs for measurements, which were compared to tables for physical development. The children's cognitive levels were evaluated, perhaps by observing the sophistication of their play. Incredibly, there were even some cases of future marriages being arranged between some babies, presumably the "best of the best."

But why should the babies have all the fun? "Fitter Families for Future Firesides" competitions came into vogue too. In one black-and-white photograph, a mother, infant perched on her lap, and presumably the baby's father sit in chairs beneath a sign that reads "Eugenics Building." Next to them hover a bevy of nurses and a doctor, stethoscope slung around his neck. A plaque proclaims: "Governors Trophy—Fittest Family." The undated picture appears to hail from the Kansas Free Fair in Topeka, where the first Fitter Families contest took place in 1920.

How best to get a message across? Teach the children. The pervasiveness of eugenics in adolescent culture was described by Steven Selden in the *Proceedings of the American Philosophical Society*:

> On a given Saturday evening in the 1920s, for example, school students could go to the movies to see the pro-euthanasia film *The Black Stork* . . . On Tuesday, the press could report on the attractive winners of "better babies contests." Sitting in class on Wednesday, these same students might open their biology textbooks to a chapter on eugenics. Finally, on Thursday and Friday, while visiting a state fair with their hygiene class, they could participate in a Fitter Families competition. If they were judged as having superior heredity, they might return home bearing a medal with a biblical inscription (Psalms 16:6), "Yea, I have a goodly heritage."

Schoolchildren heard cautionary tales about "problem families" in which criminal behavior, feeble-mindedness, and sexual excess ran rampant. In order to curtail the proliferation of problem parents and their problematic children—a function of the biblical curse, woven like a scarlet thread through generations—there was an obvious, if severe, solution: sterilization of those deemed so-cially, morally, or physically unfit.

One of the most vocal backers of eugenic sterilization in the United States was John N. Hurty, a former president of the American Public Health Association. Along with Harry Sharp, a prison physician who had vasectomized inmates at the Indiana State Reformatory, Hurty in 1907 spearheaded the first eugenic sterilization law in the nation. Dozens of other states followed suit, and the very institution that is typically considered a voice of reason—the Supreme Court—enshrined this appallingly intrusive approach to fine-tuning the gene pool. In 1927, in *Buck v. Bell*, the Court upheld Virginia's sterilization law. The subject of the case, Carrie Buck, was "represented in court by the evidence captured in a pedigree showing hereditary moral degeneracy and illicit sex, as well as mental defect reappearing through three generations of her family."

Senior Justice Oliver Wendell Holmes Jr. wrote the famous opinion for the court's 8–1 decision: "It is better for all the world, if instead of waiting to execute degenerate offspring for crime, or to let them starve for their imbecility, society can prevent those who are manifestly unfit from continuing their kind. The principle that sustains compulsory vaccination is broad enough to cover cutting the Fallopian tubes . . . Three generations of imbeciles are enough."

Paul Lombardo, a professor of law at Georgia State University who has studied this era for more than thirty years, has argued that Buck's family history was trumped up. For one thing, Buck had been raped, and her daughter, Vivian, described in the Supreme Court decision as "feeble minded," was in fact "perfectly normal." Yet the tide of legislation and the imprimatur of the Supreme Court proved influential. In the three-quarters of a century be-tween 1907, when Indiana passed the country's first eugenics law, and 1979, more than 65,000 sterilization surgeries were performed in the United States. Officials as revered as Theodore Roosevelt went on the record as saying that "society has no business to permit degenerates to reproduce their kind."

Though statements like Roosevelt's sound outrageous today, the movement was complex. Eugenics was not just about state control over reproduction, as in the United States, or genocide, as in Nazi Germany, Lombardo notes; he argues that it was an aspirational endeavor, as characterized by the slogan of the 1921 Second International Congress of Eugenics: "the self direction of human evolution."

" 'Eugenics' is like a bomb-thrower's word," says Lombardo. "When you say 'eugenics,' the bulb that goes off over people's heads is Hitler." Lombardo claims that eugenics did not have such a sinister connotation for most people. "It meant very simply picking the right mate and having healthy babies. It wasn't necessarily terrible. It was an attempt to say that we have ways of understanding how bad things happen and how babies are born who are really sick and have little hope in life because they will die young or they will struggle socially."

It's that desire for a better, happier ending that made eugenics attractive. Try this experiment: ask a random sampling of people whether it's good, bad, or inconsequential for a baby to be born blind. Most people choose the middle option. In fact, in Lombardo's experience, there's a long list of cognitive and physical disabilities that most people find scary or "bad." When he used to lecture on genetics, he'd give a quiz at the beginning. "Here are ten different conditions," he'd say. "Rank them in order of how much you'd like to avoid them." (Most feared, especially among physician audiences, was intellectual disability.) "If I did a survey of most people I know," says Lombardo, "of people who are compassionate and not mean-spirited, most people would say there are some conditions that children really shouldn't be born with if there is a way that we could avoid this."

Even Kennedy admits that had there been an accepted treatment for achondroplasia when his daughter was born, he and his wife would have gone that route "without hesitation," just as they would have for any other medical condition. "Genetic diversity may be what's best for the human race, but giving Becky the gift of health and normality—of a life free of backaches or even paralysis, of respiratory problems, of hearing loss, of orthopedic surgery . . . hell, of discrimination, of being stared at, pointed to, snickered about—well, what parent *wouldn't* do that?"

In the United States, forced sterilizations reached their peak in the 1950s, gradually declining as researchers began to question the eugenic premise that biology is the be-all, end-all. Genes are just part of what makes a person tick—a truth acknowledged in 2001 by Victor McCusick, widely considered the father of the field of medical genetics: "There are some who may view the genome in a determinist way, believing that the human condition will ultimately be seen entirely as the manifestation of sequence information and computation. We do not subscribe to such a view."

It is now fairly unthinkable to propose preventing someone, be she poor, intellectually disabled, or mentally ill, from having children or, worse, from experiencing life itself. But the specter of eugenics has not disappeared. It's just that the conversation has evolved. It's critical to remember, of course, that the forced sterilizations of the twentieth century were coerced; the people being operated on could not say no. Genetic technologies that allow people to make choices about their children are optional. No one is forcing anyone to make expensive babies via IVF, then make those expensive babies even pricier by pursuing PGD. It's voluntary, but only for those who can afford it in the first place.

Indeed, one could say the ethical issue is not whether to use embryo-selection technology, but how to expand access to it. Genetic technologies such as IVF and PGD are costly. But genetic mutations don't discriminate between rich and poor. A well-off woman with a breast cancer mutation has the option to follow in Deena Kobell's footsteps and enlist technology to have a child without passing on a legacy of heightened cancer risk. A poor woman doesn't have that choice; she must simply take her chances.

At its heart, the goal of contemporary eugenics—if that's the right term—is the relief of human suffering through the reduction of the incidence of disease. The tools are prenatal testing and genetic technologies such as PGD, genetic advances that can help ensure that babies are more likely to be born healthy. The salient difference between eugenics of yore and its modern permutation, so to speak, is the absence of institutional coercion. Today, it's individuals, not government officials, who are making the choices about which technologies to use to influence who will be born.

Alexandra Minna Stern, a professor of obstetrics and gynecology at the University of Michigan and author of *Eugenic Nation*, offers one way to differentiate between the old and the new eugenics. At a panel assembled by the Center for Genetics and Society to discuss the quest for human perfection, Stern cut right to the chase. "What we had in the past was eugenics with a capital 'E,' " she said. "What we experience today is more eugenics with a lowercase 'e.' "

Eugenics with a little "e" can be every bit as controversial as its capitalized predecessor. Everyone's personal calculus is different. One person's "designer baby" is another person's chance to rewrite a family's devastating history of disease. We are fortunate to live in an era when technology brings more people a choice—and responsibility.

As testing grows more sophisticated, the bar for a healthy baby continues to climb. It's not just about using technology to prevent babies from being born with an increased risk of breast cancer or an extra chromosome; some parents also dream of imbuing their children with "positive" traits—the perfect height, say, or perfect pitch. But concern about women choosing abortion willy-nilly because a baby has the "wrong" eye color is ridiculous, even patronizing, says Ellen Wright Clayton, a pediatrician and cofounder of the Center for Biomedical Ethics and Society at Vanderbilt University. "It offends me to my core that people believe that women have an easy time deciding to terminate a pregnancy. Having an abortion is a big deal. And aborting a wanted pregnancy is really a big deal. I think most women regard it as such."

3

The Other Scarlet "A"

Abortion's Relationship to Genetic Testing

When the second purple line appeared on the white plastic wand on a March morning in 2002, I knew next to nothing about pregnancy and even less about raising a child. It was years before I'd go on to cover parenting and pediatrics, and write about sequencing children's genomes. Yet from the first days of that pregnancy, I was already enmeshed in the most cutting-edge technologies of the time, thanks to my friend Tali, whose son was due a week after mine.

Tali had recently moved to my home state of North Carolina from Israel, where nuchal translucency testing was standard. I had no idea what it was, but I figured it was important, judging by her level of outrage that this test to gauge Down syndrome risk—a combination of an ultrasound to measure the collection of fluid under the skin on the back of a fetus's neck, and a blood draw—wasn't commonly available in the United States. Within days, she told me that she'd found a doctor who was getting certified to perform the testing. He needed subjects. Tali and I volunteered.

I had signed up blithely, without seriously considering what

I'd do if the test result came back positive. I expected good news—
and, fortunately, I got it. Now, more than a decade later, nuchal
translucency is old hat. Other, more sophisticated tests have begun
to usurp what was lauded as the latest in prenatal technology in the
early aughts.

Nuchal translucency offered both Tali and me reassurance. But
the various permutations of prenatal screening and testing do not
always provide comfort. I am witness to that in the heart of midtown Manhattan, not far from Radio City Music Hall, in a clinic
where a woman and her husband have just made a life-altering decision. The woman is forty years old, with high cheekbones and
skin the color of toasted almonds. She is twelve and a half weeks
pregnant with one baby, but minutes earlier, before she sat down
with me in an empty exam room, she was pregnant with two.

Her journey to motherhood had not been easy. The twins had
been conceived via IVF after the woman and her partner had spent
more than a year trying to get pregnant the old-fashioned way. A
week before, she had had a microarray analysis that peered deeply
into the genetic makeup of her twins. As the woman explained,
"We made use of technology throughout this process, so it would
be a shame not to take advantage of this [test]. I wanted to make
sure that, given my age, there was nothing wrong."

She knew that the microarray would reveal all sorts of genetic
blips, DNA duplications and deletions too tiny to be seen under a
microscope, some of which are associated with worrisome conditions and others of which aren't understood. The test would also
detect major chromosomal problems, of which Down syndrome is
the most common.

Even at age forty, just 1 of 100 pregnancies results in Down
syndrome. And yet the couple beat those odds: one of the twins was
confirmed to have the extra twenty-first chromosome that causes
the condition.

"You don't think it's going to happen to you and then here it
is. I still can't get over the fact. Today we reduced the baby with

Down syndrome," the woman tells me, using a common euphe-
mism for terminating one or more fetuses in cases where a woman
is carrying more than she intends to deliver. Many doctors call
this "fetal reduction." She reflects on her decision, made possible
by these new tests, as she lowers herself onto an exam table to rest.
"I look at this as a sign from God. My mother believes in karma.
I think this baby was only meant to be for twelve weeks and his
suffering was shortened," she says. She raises herself up on her
elbows and looks at her husband. "Then I feel like, 'Oh my god,
I just killed a baby.' "

Considering that women have been getting pregnant for a
very long time, prenatal diagnosis—the ability to peer inside the
womb and emerge with a snapshot of fetal health—is a fairly recent development, a convergence of medical technologies such as
amniocentesis and ultrasound with emerging insights about genes
and chromosomes. But it's the legalization of abortion in 1973 that
really served as a catalyst for change. After all, without the ability
to choose whether or not to continue a pregnancy, knowledge
gleaned from prenatal diagnosis would have remained largely theoretical. With the decriminalization of abortion, what to do became a choice.

While there are women who'd never opt for an abortion, it's
disingenuous to ignore the fact that terminating a pregnancy is one
possible outcome of earlier, more sophisticated genetic tests. The
issue of how people feel about disability and, in turn, how that
impacts their decisions regarding abortion is an essential aspect of
any discussion about advances in prenatal testing.

Yet abortion remains the elephant in the room when it comes
to prenatal testing. When I discuss my work with colleagues and
friends interested in the subject, some say, "You're not going to
mention abortion, are you? My gut tells me that I think you're
walking into a minefield if that becomes a major part of the book."
Others say, "Abortion should definitely be a chapter. How could it
not be?"

Much of the prenatal testing conversation centers on Down syndrome because the condition is so well-known, unlike others that affect far fewer people. One of every 792 babies born in the United States has Down syndrome. Compared to many other chromosomal conditions, however, Down syndrome is considered a relatively mild genetic complex. Chromosome 21 is the smallest chromosome, so the extra genetic material that accompanies a third copy is not as massive or overwhelming as it would be had it occurred on another, larger chromosome. The genetic disorder that results from a triplication of any chromosome is called a trisomy. A trisomy 22 baby, for example, probably would not make it to birth.

Starting in the 1970s, various epidemiologists began making the case that standardizing testing for Down syndrome was a public health priority. Since then, screening for Down syndrome has become broadly accepted by the medical community and, in turn, by many pregnant women and their partners. In 2007, the American College of Obstetrics and Gynecology (ACOG) expanded its prenatal screening recommendations to offer all women, regardless of age, the option of screening and diagnosis for genetic conditions, including Down syndrome.

One of the consequences is clear. In 2015, Brian Skotko, who codirects the Down Syndrome Program at Massachusetts General Hospital, published a comprehensive look at Down syndrome live-birth rates in the United States. Between 2006 and 2010, he and his colleagues calculated that 30 percent fewer babies with Down syndrome were born than were expected, due to elective terminations.

Decisions about whether to have a baby with Down syndrome tend to vary geographically and by level of education. In the 2015 study, abortions for reasons of Down syndrome were highest in the Northeast and Hawaii and lowest in the South. Asians were the most likely to terminate due to Down syndrome, while Hispanics and American Indians were the least likely.

Various forms of prenatal testing have been around for decades, but when noninvasive prenatal screening (NIPS) debuted in 2011, its greater accuracy combined with its ease of use contributed to its rapid uptake. In a few short years, NIPS, also called cell-free DNA screening, has become pervasive in the prenatal-testing market. Rather than face off with a long needle or catheter guided through the cervix or abdomen late in the first trimester, or a long needle in the abdomen in the second trimester, a quick venipuncture can collect enough blood midway through the first trimester to gauge whether the fetus's chromosomes are intact, with high accuracy and no in utero assault. Within a few weeks of a woman learning she's pregnant, her blood contains fragments of fetal DNA (NIPS actually detects DNA from the placenta, considered a proxy for fetal DNA, that is free-floating in the mom's bloodstream). The amount of cell-free DNA from the fetus and mother can then be analyzed to predict Down syndrome (and an increasing number of other chromosomal conditions) with up to 99 percent accuracy—though the concept of accuracy itself is nuanced and complex and fluctuates depending on the age of the mother. NIPS, being a blood test, also sidesteps the very small but still scary risk of miscarriage that accompanies CVS or amnio.

Initially reserved for women over thirty-five, NIPS has now spread to younger women as well, and has spawned a $500 million industry expected to balloon to $2 billion by 2020. But who gets the testing ranges widely, depending upon who goes to the doctor in the first place. Lower-income women, due to lack of access, don't seek out prenatal care nearly as regularly as more well-to-do mothers. If they do, they're often too far along in their pregnancies to get screened. Due to geographic discrepancies in Medicaid coverage, NIPS or other tests may not be covered.

What's really significant is that the decline in births of babies with Down syndrome has preceded the advent of NIPS. It would be reasonable to expect this trend to continue, even to increase, now that testing is less invasive and thus more common. Yet a small study from Eastern Virginia Medical School published in the journal *Prenatal Diagnosis* concluded that the new screening hasn't altered outcomes much. Between 2003 and 2011—the eight years preceding the introduction of the first commercially marketed non-invasive prenatal screen—the area of southeastern Virginia near the medical school (incidentally, home to the Jones Institute for Reproductive Medicine, which is responsible for the birth of the first IVF baby in the United States) welcomed about twenty babies with Down syndrome each year. In the three years after the introduction of the new screening, about nineteen babies were born each year with an extra twenty-first chromosome. In other words, a diagnosis is not necessarily a de facto fast track to abortion.

Despite its high degree of accuracy, NIPS is not perfect. Nor does it equate with a diagnosis. NIPS is a screening test; it can be complicated by a lower-than-expected fraction of fetal DNA and even by an underlying maternal cancer diagnosis. Only CVS or amnio can offer confirmation. But the message is not always getting across to women—or their doctors. Cases have been reported of women coming close to terminating pregnancies they believed were affected based on NIPS results—only to learn that they were not. Experts blame the companies that market the tests for robust advertising that they say misleads patients—and some physicians—into believing that the results are equivalent to a diagnosis. To address misunderstanding, ACOG issued a statement in 2015 stressing that any positive results need to be confirmed via other tests such as amniocentesis. In other words, ACOG emphasizes, a decision to have an abortion should not be based solely on the results of NIPS.

Yet there has been little public conversation about widespread prenatal screening and the "consequences of the transformation of every fetus—and not only the precious fetus produced thanks to complex technological interventions—into an 'at risk' entity, extensively tested, measured and evaluated by health professionals," wrote the science historian Ilana Löwy in a paper about prenatal diagnosis.

In an op-ed in *The New York Times*, "The T.M.I. Pregnancy," Patricia Volk lamented that all the testing surrounding her daughter-in-law's supposedly "normal" pregnancy had left them both feeling "guardedly happy." She recounted a series of scary ultrasound findings that turned out to be nothing, and mused: "Prenatal science has helped a lot [of] people and people-to-be. But just because a patient can know something, must she? Odds are in this baby's favor, yet every sonogram adds something scary to the pot. What is one of the most joyous times of life has turned into something ominous and fraught, loaded with the potential to go wrong."

Yet one person's anxiety is another's sigh of relief. The debate over what testing and how much hinges on so many factors. In fact, two letters to the editor in response to "The T.M.I. Pregnancy" highlight why this push-and-pull is one of the great medical and social conundrums of our time. In one, Alastair Pullen describes his experience declining all testing during his wife's first pregnancy "for all of the reasons this article mentions." Halfway through the pregnancy, he and his wife agreed to an ultrasound and discovered their daughter had a fatal condition and would not survive long after birth. "Faced with a horrible decision," Pullen writes, "we decided to induce preterm labor. Becket was stillborn. The only thing worse would have been if we had had no knowledge of her condition." Pullen had first decided not to test but ended up grateful he changed his mind. He and his wife welcomed testing in later pregnancies; they now have three healthy children, and, he says, "the barrage of testing affirmed our excitement."

Ingrid Chafee, on the other hand, gave birth when no tests were available. She was shocked when she delivered her firstborn in 1965, only to learn he had hydrocephalus and spina bifida. Surgery repaired much of the damage, but her son—who now holds a doctorate from Oxford—still has physical problems. She concludes: "He has said many times that he is glad that there were no ultrasound tests available at the time of his birth. If there had been, he wouldn't be here. To know or not to know? It's up to each to decide."

□

"Collective fiction" is a term that's been applied to our cultural hesitancy to talk about abortion as it relates to prenatal testing. The phrase was used by researchers at the University of California, Los Angeles (UCLA), in the journal *Fetal Diagnosis and Therapy* in the early 1990s, a time when the U.S. Department of Health and Human Services had indicated its intention to offer prenatal screening and counseling to at least 90 percent of pregnant women. In their paper, Nancy Anne Press and Carole Browner of UCLA's Mental Retardation Research Center specifically explored why pregnant women in California chose to accept or reject the maternal alpha-fetoprotein (AFP) blood tests that predated today's non-invasive prenatal screening tests. (Elevated AFP is still an excellent indicator of neural tube defects such as spina bifida, which isn't detected by NIPS.) Press and Browner, medical anthropologists, argued that pregnant women and their physicians created a "collective fiction" that incorrectly placed the testing "within the domain of routine prenatal care and denied its central connection to selective abortion and its eugenic implications." This allowed them to gloss over contentious ethical and moral issues and normalize the decision to seek testing.

More than two decades later, prenatal screening options have expanded, but the collective fiction remains entrenched, says Megan Allyse, a bioethicist at the Mayo Clinic who studies the ethics of NIPS. "These are all euphemisms, that this testing is really just about information and that we're not talking about abortion, although we really are," says Allyse. "That is even the narrative in online forums where women talk about whether they should do the testing. Other women will respond: 'You have to do everything you can do to ensure you have a healthy baby.'"

Perhaps we take refuge in circumlocution because it feels strange to acknowledge that prenatal testing allows us to play a role in deciding what sort of child we will have. It by no means rules out every possible problem, of course. There are scores of issues that aren't tested for or can't be detected during pregnancy. But the blood tests and ultrasounds and needles in the belly are designed to offer the best available snapshot of fetal health in real time. In many cases, such as the routine ultrasound that revealed a cyst on my daughter's brain that could have indicated trisomy 18 but ended up fading away on its own, those snapshots turn out to be unnecessarily alarming.

Say your car is due for an oil change. While you wait, imagine the service manager comes over to suggest a new air filter and, as long as you're there, a tire rotation too. A savvy consumer doesn't automatically say yes.

The automotive analogy is one that the genetic counselor Mary-Frances Garber falls back on frequently. She also likes to compare the array of prenatal tests now available to the bountiful choices on a Chinese restaurant menu. Garber counsels couples at a hospital in Newton, Massachusetts; in her private practice, she offers private counseling and support groups that help couples navigate their emotions in the wake of prenatal diagnoses. With more testing offered, Garber identified a need to help couples process the aftermath of a pregnancy gone awry. One of her groups focuses on patients and their partners who have ended their pregnancies following a prenatal diagnosis for a variety of conditions. The couples in the circle have different experiences but share a common thread: they each were faced with a gut-wrenching decision.

"In the hospital, we don't get to focus on their grief or bereavement," she says. "That's a luxury. With all the different things we can offer testing for, there's a need to help people adjust and heal and focus on their emotional response." She observes all sorts of reactions, running the gamut from typical grief—the expected mourning that follows the discovery that your healthy baby narrative is not to be—to pathological grief, which paralyzes and doesn't loosen its grip. Many couples also experience marriage problems because of nonparallel grieving, in which the two process their loss differently. "I tell them that grief is okay in this room, but I really want to get rid of guilt," says Garber.

There is often a default expectation on the part of medical providers that pregnant women will want prenatal testing. Over and over, Garber hears the same thing: so many people say they didn't fully understand what they were signing up for when they agreed to do genetic testing. They say yes to one test, which leads to another, and another. Women who thrive on lots of information welcome the array of choices, while others may feel threatened. If asked, many couples will say they're testing for peace of mind, because they want a negative result. Rarely have they given serious thought to what they'll do should the results come back positive.

Genetic counselors can offer clarity about the various options, but they are rarely involved at the initial stages. Genetic counselors tend to cluster at major medical centers, which can afford them, and not at ob-gyn offices, where most women receive care. The average ob-gyn's office doesn't employ genetic counselors, because insurance companies don't regularly reimburse for their services. Those private ob-gyn offices that buck the trend often have genetic counselors who are salaried employees of the laboratories that provide the tests used in that clinic, and this presents a potential conflict of interest. The result of inconsistent access to counseling, Garber suspects, is that patients are nowhere near as informed as they could be. "We don't want to make pregnancy a disease," says Garber, who says she is pro-choice. "Some of these patients are on a treadmill and they can't get off. Once you give blood, you are on that treadmill."

Recently, Garber counseled a couple who couldn't decide what to do. One-half of the couple was inclined to make a decision based on fear and anxiety: worry about the child not being able to live independently, worry about having to quit work to care for the child, worry about the expense associated with the child's disabilities; the other half carried the ultrasound photo everywhere and wanted to make a decision of the heart. Garber challenged each of them to approach the decision from their partner's perspective— to look at the ultrasound picture, to explore the fears—and they were able to reach consensus. They decided to continue the pregnancy. "I remind people that every time we have a baby we are hoping for a typical child," says Garber. "But we can be blindsided. There is an unknown with every child."

That was certainly the case with Ryan Docherty. When I met Ryan, he had just celebrated his first birthday, and the helium-filled Mickey Mouse and Elmo balloons from his party were still aloft. He started walking recently and likes to drop a toy or a spoon—anything, really—over the side of his highchair with a chirpy "Uh-oh!" and wait for an adult to retrieve it. Then he drops it again and giggles. In other words, Ryan, who has pale blue eyes, squeezable cheeks, and downy reddish-blond hair, is completely typical for his age.

"And to think we almost terminated him," says his father, Steven Docherty, who has cut right to the chase. "He's a miracle baby."

The ethics of abortion are set to become much more complicated as more women have access to powerful genetic tests such as microarray, for these tests can identify genetic flaws that are not readily understood. Microarray had confirmed that the almond-skinned woman who had the fetal reduction was carrying one twin with Down syndrome. But in the case of Ryan's more ambiguous genetic errors, confirmation was the easy part. It was the interpretation—figuring out the significance of the problems that microarray had detected in utero—that proved difficult.

When she was pregnant, Ryan's mom, Jen Sipress, had a microarray test. You'll recall that chromosomal microarray analysis can detect deletions and duplications of genetic material—errors that are far smaller than an entire extra chromosome. But just because they're smaller doesn't mean they can't wreak havoc. Some are associated with genetic disorders; many more aren't associated with anything because they're so newly discovered or because they don't appear to be detrimental according to the limited amount of research that exists. Sipress, forty-two, is a New York City narcotics prosecutor; she thrives on evidence. When her test results came back, the evidence was disconcerting: Ryan, still in utero, had not one but two findings—"variants of uncertain significance"—inherited from his mother and his father. Docherty had passed down a duplication involving six genes, while Sipress had contributed a deletion on chromosome 15 involving four genes. In general, deletions are considered more worrisome than duplications; our bodies can often deal with some extra genetic material, but it's not as easy to compensate for DNA gone AWOL. To make matters worse, one of the four missing genes had been associated in the medical literature with intellectual and developmental delay. Here's where things got really confusing: Sipress was missing that same gene and she didn't appear to be affected at all. She worked hard as the family's primary income earner, putting drug dealers behind bars. She hadn't even known she was missing any genes until the microarray results came back. But genes—or their absence—can affect people differently; it's a phenomenon called "variable expressivity."

Before the amniocentesis to collect fetal cells for the microarray analysis, Sipress and Docherty had decided that were they to learn that their unborn child wouldn't be able to live independently as an adult, they would end the pregnancy. When they got the results, they leaned toward abortion. After talking to their doctor, Ron Wapner—the same physician who had tended to the mom pregnant with twins, and author of a *New England Journal of Medicine* study about microarray's effectiveness—they changed their minds. As Sipress recalls, Wapner said, " 'I get people coming in here who . . . want to know this is 100 percent fine.' And he said, 'I can't give you 100 percent. I can give you 80 percent.' And I said, 'I'm going to take those odds.' "

Emotionally, it was a terrible time for Sipress and Docherty. Ryan was their first child, and he had been conceived after two rounds of IVF. But Sipress doesn't regret finding out. "I don't understand why even women in their twenties aren't undergoing this testing," she says. "Knowledge is power. Doesn't everyone realize that?"

It's certainly made for some awkward conversations with her husband's family in Scotland, who know about the missing genes. "They ask if there is something wrong with the kid, and I say, 'Technically, yes, but he's not exhibiting any symptoms,'" says Sipress. To that end, Docherty, who stays home with Ryan, is a vigilant observer. "Are we still worried?" says Docherty. "Absolutely." It's easy to attribute every behavioral challenge—Ryan's not a good sleeper, but neither are lots of babies—to the missing genes. Anticipating this, Wapner has cautioned them against engaging in this sort of genetic determinism. "He said, 'Go about your business. If you feel something is really wrong, then you act.' To be honest," says Docherty, "Ryan doesn't have a problem, as far as I can see."

To what end are we willing to go to detect disability? Once we find it, is there a dividing line between "good," or tolerable, disabilities and "bad," or intolerable, limitations? How do we decide which ones may warrant abortion and which are acceptable? What feels overwhelming to one person—the birth of a child with a genetic disorder—may feel like God's gift to another. Who are we to judge what—who, more accurately—is a gift and who is a burden?

All things being equal, it's tough to argue that having a significant disability is a good thing for a particular person. As Paul Lombardo discovered in the previous chapter, most people, given the choice, would rather not have a disability of any kind, nor would they want their children to have a disability. "There is nothing wrong with saying, 'I'd rather my child not have this disability,' as long as we don't go from there to the view that people with disabilities are worth less," says Ruth Faden, former executive director of the Berman Institute of Bioethics at Johns Hopkins University.

One of the best-known disability researchers is the late

bioethicist Adrienne Asch, who herself was blind from her first weeks of life. She was strident in her opposition to prenatal testing and to performing abortions specifically to avoid having children with disabilities. At the same time, she believed strongly in reproductive rights and access to abortion.

Although the two positions appear at odds with each other, Asch didn't see it that way. It is okay, she believed, to pursue abortion if you don't want to have a child; it is not okay to end a pregnancy because you don't want *that* child. In other words, it's fine to have an abortion because you don't want to be a parent; what's not acceptable is to have an abortion because you don't want to parent a disabled child.

"If public health espouses goals of social justice and equality for people with disabilities, as it has worked to improve the status of women, gays and lesbians, and members of racial and ethnic minorities, it should reconsider whether it wishes to continue endorsing the technology of prenatal diagnosis," she wrote in the *American Journal of Public Health* in 1999.

The psychiatrist Dorothy Wertz had a wry take on that view. Wertz, who worked at the Eunice Kennedy Shriver Center in Waltham, Massachusetts, had researched attitudes about disability among genetics professionals, doctors, and patients. "To answer Adrienne Asch's critique of genetic screening, Dr. Wertz joked that most abortions already take place for genetic reasons: half of the fetus's genes are from a man with whom the woman does not wish to have a baby," wrote Dan Kennedy in his memoir about his daughter's dwarfism. "From there, it is a short leap to choosing abortion for genetic reasons such as serious illness, not-so-serious illness, or conditions that simply don't match the prospective parents' vision of the perfect child."

❐

Setting aside the politics of abortion, it's instructive to examine how disability is perceived in our society through the lens of what's known as "wrongful birth." In 2012, a Portland, Oregon, couple was awarded nearly $3 million from Legacy Health System, where the woman's prenatal testing was performed, after their daughter, Kalanit, was born with Down syndrome. Deborah and Ariel Levy sued for damages, alleging the physician who had performed the CVS procedure had mistakenly removed maternal tissue instead of fetal tissue. Had they known that Kalanit would be disabled, they said, they would have had an abortion. Nevertheless, they said they loved their daughter "deeply" and would use the money for her lifelong care. When the verdict was announced, Deborah Levy began to cry. "One juror visibly held back tears," reported *The Oregonian*. "Another wished them peace."

"Wrongful birth" lawsuits are allowed in more than half the states, but they aren't filed often—a few times a year in the United States, experts estimate. Why so few? Parents must go on record as saying they would have aborted their child had their doctor told them there was a problem. That's a tall moral order, especially in a society that invites online castigation. One blogger wrote of the Levys, who have two older sons: "First, how can any parent simultaneously claim to 'love' their child, yet also wish that they had aborted them? . . . They just had the perfect family until their little girl had the audacity to be born with an extra chromosome, and now, they have to be paid off in order to deal with the burden of raising her."

How we view disability varies widely depending on the situation. We are a society rife with contradiction. Just because millions of people cheer Target for featuring children with disabilities in their ad campaigns doesn't mean that they'd want to raise a child with disabilities—if they were given the choice. The opportunity to choose whether to have testing in the first place and then what to do about the results is arguably the most significant outcome of increasingly powerful genetic tests. It's not just about whether or not to have a child, but about which sort of child to have.

Addie Morfoot, a writer in Brooklyn, says she "didn't know a thing" about genetic testing when she was newly pregnant in 2010. At her obstetrician's office, a nurse recommended she be tested to see if she was at risk for transmitting a defective gene to her child. Morfoot took her advice. "It didn't really register at the time, all the implications," says Morfoot. Cystic fibrosis was one of the diseases she was screened for, even though no one in her family had it.

Morfoot learned that she and her husband both were carriers for a particularly aggressive mutation, meaning they were unaffected but stood a one-in-four chance of having a baby with the disease. They were among the more than 10 million Americans who don't know they're carriers for cystic fibrosis. Median survival is 37.5 years, but Morfoot was told that a baby who inherited her particular mutation would be lucky to make it out of childhood.

Morfoot opted for an amniocentesis, which confirmed that her baby—a girl whom they'd named "Annie"—had inherited two defective copies of the gene. She cried and deliberated, but deep down, she knew all along what she would do. Abortion, for Morfoot and her husband, was the only humane option. "I just felt that it was too much to bring a child into this world who would suffer like that," says Morfoot.

At the abortion clinic, she felt compelled to tell the doctor this wasn't a matter of not wanting a baby. Morfoot wanted to differentiate her choice to abort a much-desired child with a life-limiting condition from the decision-making process associated with an unwanted pregnancy. "The baby is sick," Morfoot told her. To me, she says: "When you're getting an abortion, you think that people think you don't want the child. But we wanted the child. I felt so alone."

She struggled to find her bearings in a world that felt completely foreign. While she is grateful for the existence of carrier screening, she doesn't think that she was fully informed about what she was signing up for. She recalls a nurse suggesting she get the test done and handing her a pamphlet with information, and she recalls moving ahead since she'd already had the necessary blood drawn for other reasons. "I just agreed without really knowing anything about these tests," she says.

Now, in retrospect, Morfoot wonders why insurance typically covers carrier screening but doesn't often cover the technology that can address a couple's shared carrier status—preimplantation genetic diagnosis (PGD), which, as we've seen, permits parents to select an embryo that is free of certain genetic conditions. This was how Deena Kobell conceived Eve, who does not share her mother's breast cancer mutation.

When Morfoot shared her experience in the online magazine *Salon*, she was trounced in the comments section, often by parents of kids with cystic fibrosis or adults living with the disease—or both. Morfoot and her husband found the empathy they were seeking at a support group along the lines of those that Garber facilitates. In a room of couples who had terminated a wanted pregnancy after genetic testing had revealed a problem, they took solace in others' narratives of pain. And they were surprised by how many other parents had found themselves in a similar situation.

More prenatal testing, of course, means that more expectant couples stand to learn that their pregnancy is troubled. At the same time, abortion in general is becoming harder to access, as some states have required abortion clinics to meet stringent standards that are designed to discourage abortion. A few states have zeroed in on the motivation behind an abortion, considering or adopting legislation that prohibits women from seeking an abortion due to disease or a genetic condition.

In 2013, North Dakota became the first state to pass a law that bans abortion for the purpose of eliminating any "genetic abnormality." The law's definition of what qualifies as an anomaly is quite broad, ensuring that nearly any genetic defect could theoretically be included. Had Morfoot lived in North Dakota, it's possible that she might not have been allowed to end her pregnancy, although it's unclear how enforceable North Dakota's law is in reality. The burden of proof is quite high to show unequivocally that a woman decided to abort solely because of a genetic anomaly. Still, many doctors are concerned, because they feel the legislation infringes upon a patient's ability to make choices and their own ability to practice medicine. "They can get in trouble if they have an inkling that a woman is having an abortion for fetal anomaly," says Elizabeth Nash, whose job it is to track this type of legislation for the Guttmacher Institute, a think tank committed to reproductive rights. "Abortion for fetal anomaly is much more out in the open because you essentially have a paper trail." Medical records, of course, would document any fetal diagnosis, but Nash points out that *Roe v. Wade* affirmed a woman's right to access abortion without requiring her to provide her motivation or reasoning. Despite that, Oklahoma, which bans abortions due to gender preference, asks providers if an abortion was provided for reasons of gender selection on the abortion reporting form that they file with the state.

In 2016, an Indiana bill echoed similar legislation in Ohio and Missouri, seeking to outlaw abortion due to disability. The Indiana bill, which was blocked by a federal judge shortly before it was slated to become law in July 2016, would have prohibited abortion on the basis of race, gender, or disability, and called for "disciplinary sanctions and civil liability for wrongful death if a person knowingly or intentionally performs a sex selective abortion or an abortion conducted because of a diagnosis or potential diagnosis of Down syndrome or any other disability." The language defined "any other disability" as "any disease, defect, or disorder that is genetically inherited." Even if the bill were to become law, it remains to be seen if a ban could actually be implemented.

Meanwhile, multiple states mandate the sort of information a woman is required to receive after a prenatal diagnosis of Down syndrome. Nash, who is the expert on which state is considering what law restricting abortion, is keenly aware of how the availability of prenatal diagnosis has transformed how people feel about raising disabled children. Before the advent of prenatal diagnosis, whether or not to have a disabled child was not a matter of choice. Now that it's possible to identify a range of genetic conditions prenatally, women face conflicting pressures. Some of those pressures come from disability rights activists who have struggled for recognition and respect from society. "If you say 'I'm not ready to take on this responsibility of raising a child with special needs,' there are people in the disability rights movement who see that as a failure," she says.

At least fourteen states prohibit abortion after 20 weeks postfertilization, although three of them include exceptions to allow abortion at this point if the fetus has a condition that is considered incompatible with life. (Determining what falls into the "incompatible" category can be tricky; Down syndrome would not qualify, for example, but trisomy 18, which is almost always fatal soon after birth if not before, would seem to meet the definition. Yet there are always outliers, children who beat the odds and survive, albeit severely disabled. In 2012, as the Pennsylvania senator Rick Santorum was chasing the GOP nomination, he took time away from the campaign trail when his three-year-old daughter, Bella, who has trisomy 18, was hospitalized.)

Late-term abortions are as rare as they are controversial. Just over 1 percent of abortions take place at 21 weeks of pregnancy—when a woman is five months along—or later. Women who learn of a complication well into the second or third trimester and decide to terminate often have to travel out of state to the handful of doctors who still perform abortions later in pregnancy.

Warren Hern is one of those doctors. His office, the Boulder Abortion Clinic, occupies a slightly bedraggled brick building the color of sand across from a Whole Foods and not far from downtown Boulder, Colorado. I am buzzed into the foyer, with its mirrored reception window so visitors cannot see inside. All the blinds are shut tight. I forfeit my driver's license before I am allowed into the lobby; a receptionist makes a copy for security purposes.

I wait for Hern, who is in the midst of doing an abortion, in a private room with four vinyl chairs, three faded Southwestern tapestries on the wall, and a container of condoms on a shelf. A sign says: "Please help yourself to free condoms." I take a red one.

Hern walks in, tall and broad-shouldered, sporting a full head of silvered hair and green scrubs. Thinking about what Garber and others have said about the overmedicalization of pregnancy, I ask Hern for his thoughts. Do Hern's patients express regret for having accepted the prenatal tests that led them to this office with its free condoms and bulletproof glass? Would they have preferred to not have been confronted with such a difficult decision, to revert to the reality of previous generations? "I haven't heard from them that they wish they had not known," says Hern. He peers down at me through largish gold-rimmed glasses of the kind that were popular in the eighties. "I am always in favor of more information," he says.

In a paper he published in *Prenatal Diagnosis*, Hern listed 160 different fetal disorders for which he has performed an abortion. He does not see it as his role to judge, and he will do an abortion up to 39 weeks as long as he believes the procedure is safe for the patient.

His patients, in many cases, did not seek early prenatal diagnosis. For some, it wasn't available. Others decided to hope for the best. Some even chose to bypass noninvasive prenatal screening, yet ultrasound later in pregnancy revealed a problem. The week that I visited, Hern had performed abortions for women from France, which requires two doctors to sign off on abortions after the first trimester, and Portugal, which allows abortion up to the tenth week of pregnancy and places various limits afterward. Most U.S. states restrict later-term abortions, but Colorado does not. "By the time people get to my office," says Hern, "virtually no one else will see them."

Through good luck and good guards, Hern marked forty years in business in 2015. (When I met Hern that year, he told me that he was planning an anniversary celebration, but he wouldn't tell me when it would be held, citing violence from anti-abortion activists. "If they know when it is," he said, "they will kill me.")

Hern is vilified by anti-abortion stalwarts, but to his patients he is a savior. He is as committed to a woman's right to choose what to do as anti-abortion supporters are committed to the opposite. "This is a decision that is up to each family," he says. "If they know they will have a child who is catastrophically impaired, they should be able to act on that."

❐

Even if abortion-for-fetal-anomaly became far more common, disability could never be fully engineered out of the human gene pool. There are people for whom abortion is not an option and others who choose not to do prenatal testing at all. While advocates worry about a campaign to eliminate people with disabilities, others advance the opposite worry: when women receive a prenatal diagnosis, they may also get literature that paints an overly rosy picture of what it's like to have a child with the condition. Just as some states specify that women can't seek abortions due to Down syndrome or other conditions, at least twelve states require that women receive written information from providers on what parents with a child who has Down syndrome might expect. The stated goal of the information is to make sure that women are better informed, but the ultimate goal appears to be to forestall a knee-jerk tendency toward termination of an affected pregnancy.

In Pennsylvania, the stipulation is known as "Chloe's Law" after a girl with Down syndrome whose father advocated for its passing. The information required to be shared with parents who receive a prenatal diagnosis is intended to be up-to-date and evidence-based—neither anti-abortion nor pro-choice but "pro-information," as Chloe's father, Kurt Kondrich, characterizes it. "As prenatal genetic testing advances," he has blogged, "I challenge people to ask the question, 'Who will be next to be identified . . . and eliminated because they don't meet cultural mandates for perfection?' "

Yet while many parents of children with Down syndrome rejoice at these laws, opponents often include state medical associations, which reject legislation that seems to interfere with the doctor-patient relationship. Some opponents believe that telling doctors they must share certain brochures smacks of other politically motivated policies aimed at pregnant women, such as requirements in some states that they undergo ultrasounds before proceeding with abortions.

Pennsylvania's Down Syndrome Prenatal and Postnatal Education Act took effect on October 1, 2014; I learned about it later that month as I was touring the Special Delivery Unit, which caters to babies with prenatal diagnoses, at the Children's Hospital of Philadelphia (CHOP). A genetic counselor handed me a copy of the pamphlet from the state Department of Health that she was now mandated to hand out to any woman receiving a prenatal diagnosis at the hospital. It features a photo of a radiant, Mohawked baby boy with Down syndrome and provides information about how the condition occurs, its characteristic features, including cognitive impairment, and other associated medical complications. It does not discuss potential choices that a woman may face following a prenatal diagnosis. A reporter writing about the pamphlet for *The Philadelphia Inquirer* wondered: "Can information be considered unbiased if it doesn't mention the abortion option?"

It is understandable that Down syndrome advocates and parents of children with Down syndrome are supportive of these laws. As births of babies with Down syndrome decrease in proportion to the population, parents of such kids are rightfully concerned. If their numbers dwindle, will research dollars follow suit? Will early-intervention services shrink? Will society become less tolerant? But the bioethicist Arthur Caplan has argued that the legislation—which includes the federal Prenatally and Postnatally Diagnosed Conditions Awareness Act—undermines the central tenet of genetic counseling: neutrality.

For decades, genetic counselors have been instructed in the art of being "nondirective"—sharing information and explaining it, but then stepping back to allow the patient to draw his or her own conclusion about what to do. A genetic counselor is not even a counselor—an adviser—in the typical sense of the word; in truth, the profession might do well to consider a name change. Genetic counselors are not supposed to give counsel as much as they are supposed to impart knowledge. Of course, the very act of presenting information involves value-laden choices. What people choose to do with that information is then up to them.

Are high rates of termination a result of genetic counselors gently guiding women toward that outcome, as advocates of the laws suspect? Or are women and their partners making that decision with their eyes wide open, concerned about having a child with mental and physical challenges? Caplan summarizes the tone of the legislation like this: "Kids with Down syndrome may have issues, the law concedes, but medical advances, devoted parenting, and societal resources can help ameliorate them." All that is true, in fact, but it is not necessarily the entire story. Should we be transforming the field of genetic counseling from an objective approach to a point of view? Caplan argues that this is just the beginning of the so-called "slippery slope." If positively normed legislation exists to champion the birth of people with Down syndrome, won't parents of other children with disabilities soon come forward, pressing for their own sunny laws?

Ricki Lewis, a genetic counselor who has authored a textbook on human genetics, echoes Caplan's concerns that requiring genetic counselors to, as she writes, "more positively spin life with trisomy 21 Down syndrome may obscure the medical and scientific facts, misleading patients." Each new edition of her book documents the continuing rise of prenatal testing and the decline in numbers of people with Down syndrome. She writes: "Most of my book's information about trisomy 21 is under the heading 'Abnormal Chromosome Number.' I can't be politically correct about that—normal means 'common type' and an extra chromosome is not common."

In the ninth edition of her textbook, she cites a Danish study that found that even before noninvasive prenatal screening was introduced, between 2000 and 2006, "the number of affected newborns was halved, those diagnosed prenatally increased by nearly a third, and the number of diagnostic tests (CVS and amniocentesis, which are more invasive than maternal blood test screening) fell by half."

In the twelfth edition, she discusses legislation such as Chloe's Law, agreeing with Caplan that "if laws compel genetic counselors to talk more about the happy healthy xylophone-banging little children, and less about the toddlers who sport scars from heart surgeries or develop leukemia, patients might leave counseling sessions with skewed views of life with an extra chromosome 21."

❐

Do you remember, from biology class, the fancy footsteps that characterize meiosis, the process of cell division that produces gametes—sperm in men, eggs in women? The genetic counselor Mary Linden is guessing the answer is no, which is why she prefaces her reproductive refresher with a lighthearted pronouncement: "It's time to go back to college," she announces to the six couples and two single women seated around a table at the vaunted Colorado Center for Reproductive Medicine (CCRM) on a summer day in 2014. A motley mix of women lugging Prada bags and others in hiking boots, plus two husbands slyly checking their phones for an update on the U.S.–Germany World Cup match, they're here at the mandatory 10:00 a.m. genetics class because they want to become parents. For a variety of reasons, things have not been going according to plan. Now they've turned to CCRM, a fertility clinic in the foothills of the snow-capped Rockies that draws patients from around the country and abroad. The attraction is not the picturesque setting; it's the success rates, which are consistently higher than those of many other clinics, even though the average patient age is forty—well into what doctors consider the red zone for getting pregnant.

"Meiosis" is a term most haven't heard since freshman biology, so Linden walks them through the nitty-gritty of cell division, the swirls of DNA replicating, dividing, and swapping fragments of genetic material. When the process runs like clockwork, chromosomes pull apart and recombine seamlessly. The end result is why we don't all look the same; this genetic give-and-take makes us unique. But much as an aging brain may be hard-pressed to recall a particular fact, an aging egg—any egg, for that matter—can't always handle the intricate choreography of meiosis. When a slip-up occurs as chromosomes reunite, an error results. The impact can be significant: chromosomal errors are the leading cause of miscarriage.

Many of the women in the class have endured a string of miscarriages, if they've managed to get pregnant at all. Because women undergoing fertility treatment are routinely offered more tests than the average pregnant woman, their experience serves as a useful guide to the dilemmas inherent in opting for prenatal screening. In the CCRM class, they learn about a technique called comprehensive chromosome screening (CCS) that analyzes embryos just a few days after embryologists combine sperm with egg in a glass petri dish. As with preimplantation genetic diagnosis, the goal is to identify healthy embryos. Only those embryos with the correct number of chromosomes are available for transfer to a woman's womb; any chromosomally abnormal ones are strictly off-limits according to clinic policy. Because chromosomally abnormal embryos are out of bounds, it's inevitable that some abortions will be averted.

It makes for an unusual reproductive dynamic: Women who get pregnant naturally can't choose to avoid chromosomal errors—"aneuploidy" is the scientific term—because they can't control which egg will combine with which sperm. But women whose babies are being conceived in vitro can by opting for CCS. At CCRM, 90 percent choose to do CCS, in which a few cells are carefully siphoned from a days-old embryo and analyzed for normal, or euploid, chromosome count. "We will find some aneuploid embryos for all of you," Linden announces, which results in some raised eyebrows of surprise. "Only euploid embryos are transferred at CCRM."

In plain speak, that means that women who opt for chromosome screening, assuming they get pregnant, are pretty much guaranteed a child with no major chromosomal problems. An embryo with Down syndrome, for example, would be discarded. There are no extra or missing chromosomes at this clinic; embryos with the wrong number or arrangement of chromosomes—the doctors here have seen errors in every single chromosome, errors so devastating that such embryos wouldn't even make it past the first trimester—are rejected. As a result, CCRM—and many other IVF clinics where CCS and similar genetic analyses are growing in popularity—sidesteps the potential for abortion by refusing to even transfer chromosomally abnormal embryos. In doing so, it navigates other ethical thickets, acting as arbiter of what sort of embryos should be given the opportunity to become a child. This sort of vetting is the "height of hubris," says Mark Leach, an attorney who has both a daughter with Down syndrome and a master's in bioethics. "It's your life and you should get to decide."

It strikes me as an ethically fraught choice to decide that only chromosomally normal embryos should get a shot at becoming babies; what if, say, you have a brother with Down syndrome whom you adore and you'd be happy to welcome a child like him? But what some would consider eugenics, William Schoolcraft, CCRM's founder, considers best practices. He's in the business of making healthy babies. By selecting only chromosomally ideal embryos for use, he slashes the risk of miscarriage, which is a mark of failure at any fertility clinic. Herein lies a key distinction: while PGD is typically performed to eliminate disease that runs in a family, CCS has a different stated motivation—improving the chances of a live birth, which is good for the parents and good for the clinic's outcomes.

Research has shown that CCS lowers miscarriage rates and

increases live-birth rates. "There is no controversy about doing this because we all know that aneuploidy contributes to pregnancy failure," says Mandy Katz-Jaffe, who developed the specific CCS technique used at CCRM. "Our goal is a healthy, chromosomally normal live birth."

Put that way, who wouldn't want the chance to ensure a healthy baby? "You start crossing a threshold," says Jason Elliott, who has come to CCRM with his wife, Katie, to do IVF coupled with CCS. "We thought, 'Okay, we're just going to surrender to the science.'"

As a parent, would you take advantage of genetic technologies to conceive a baby who doesn't have Down syndrome? Science is making progress in developing treatment for people with the condition. If you can treat a condition with various degrees of success, do you still need to prevent it? In Boston, where much of the work creating therapies is taking place, it's a question that parents and researchers are asking.

4

Silencing a Gene

The Future of Down Syndrome

On July 16, 2013, a day before the cell biologist Jeanne Lawrence made headlines across the world, she sent two important emails. One landed in the inbox of the family of Melissa Reilly, a woman in her twenties with Down syndrome. Lawrence had previously invited Reilly to speak to her first-year medical students about what it was like to have three copies of a particular chromosome in every cell in your body when there should be just two. The other email was addressed to the Massachusetts Down Syndrome Congress (MDSC), a parent network that is among the country's most active advocates for people with an extra twenty-first chromosome. Months earlier, in casual conversation, Lawrence had told Maureen Gallagher, executive director of MDSC, that she was involved in Down syndrome research. Such a revelation was hardly unusual coming from a scientist at a major academic center such as the University of Massachusetts Medical School in Worcester, where Lawrence works. As a courtesy, Lawrence was giving Reilly and Gallagher advance notice of her latest study.

It was no usual study, however.

The Boston area, on July 17, 2013, was uncharacteristically

steamy, and as the day wore on, the 95-degree heat felt like an apt metaphor for the buzz filtering through the Down syndrome community. At 1:00 p.m. Eastern time, the prestigious science journal *Nature* had lifted its embargo on the news that Lawrence and her colleagues at UMass had managed to silence the extra copy of chromosome 21 that characterizes Down syndrome. The achievement unspooled in a cell in a petri dish, not in a person, but it was considered a major first step in potentially treating Down syndrome.

Down syndrome, as we've seen, is a disorder of surplus genetic material. Too many genes make it hard for cells to regulate their protein production. This surplus results in intellectual disability and physical delays, plus heart defects in about half of affected children, some of whom will require surgery. People with Down syndrome are also at greater risk for leukemia and Alzheimer's, among other diseases. But in the past thirty years, there's been a seismic shift in how people with Down syndrome are treated by the medical profession. Early intervention and advances in medical care have vastly improved the quality of life for people with Down syndrome, who used to die in their twenties, usually as a result of untreated congenital heart conditions, but now routinely live into their sixties. All this is to say that for people with Down syndrome, life is certainly better than it ever has been. Yet now here was Lawrence, suggesting that the extra chromosome associated with Down syndrome could be shut off, like a light switch. Silenced.

Gallagher, scrambling to understand the impact, called Brian Skotko, who calculated the decline in Down syndrome births referenced in the previous chapter. Skotko codirects the Down Syndrome Program at Massachusetts General Hospital but wasn't involved in the research. His own sister has Down syndrome, and he has authored numerous papers showing that relatives value their family members with the condition. "I asked Brian, 'What do you think the implications are for this research?'" Gallagher says. "He said, 'I think it's one of the most significant advances in

Down syndrome research since it was discovered that Down syndrome is caused by an extra twenty-first chromosome.'"

Lawrence's research showed that when a gene was inserted to "hush" the third copy of the twenty-first chromosome, brain cells showed impressive growth. Her work involved tinkering with human cells in the confines of her laboratory, but her goal is to intervene at conception, during pregnancy, or after birth. Lawrence, as a responsible researcher, wasn't trumpeting a cure for Down syndrome; she made it clear that her discovery could potentially allow those with the syndrome to sidestep some of their developmental problems, but was unlikely to reverse the biology that underpins the condition. Still, the chatter among scientists and parents of kids with Down syndrome was loud and insistent and questioning. It was so loud that Maureen Gallagher knew she had to do something, issue a proclamation, calm the waters. So she sent the MDSC's 4,000 members a letter acknowledging what everyone was thinking, that "although it is an exciting discovery, it will bring with it many ethical and emotional issues." Chief among those is whether parents of kids with Down syndrome would actually want to shut off that extra chromosome.

To further complicate matters, Lawrence's discovery and Gallagher's public relations offensive were of course unfolding against a backdrop of the increasingly sophisticated and controversial prenatal tests that have become available lately. Noninvasive prenatal screening can indicate Down syndrome in utero in the first trimester, a good six weeks earlier than previous-generation tests that measured levels of proteins and hormones in the mother's blood.

Earlier detection, of course, allows for earlier treatment—assuming treatment is available or desired—or earlier abortions. "Most families are looking for opportunities to help their child with Down syndrome lead fulfilling and productive lives, then all of a sudden there are these scientific advances that are making society question the value of their children's lives," Gallagher told me a few days after writing the letter to her members. "We have some families who, after their babies are born, are looking for treatments and therapies. Then we have people who wouldn't want to change their children for the world and aren't looking for a cure. We tried to be sensitive to both sides."

Responses to Gallagher's letter were both grateful and confused. Some parents thought that the breakthrough seemed like science fiction; others rejected the idea of changing their kids; still others were intrigued and wanted to know more. "It brings up a lot of emotions and some sadness on the part of parents because the future of Down syndrome is changing," Gallagher says.

□

In 1961, long before "political correctness" had joined the universal lexicon, several renowned genetics experts wrote to the British medical journal *The Lancet* requesting that the name used to describe having three copies of chromosome 21—"Mongolian idiocy"—be changed; Down syndrome was one of several suggestions put forth, after the English physician John Langdon Down, who was the first to accurately describe and classify the condition using photographs and measurements of the heads and facial features of people with the syndrome. Mongolians, argued the experts, were no more genetically likely to have an extra chromosome than other populations. It didn't hurt that the Mongolian government had expressed its pique to the Director General of the World Health Organization. The WHO acquiesced.

Although Down syndrome bears John Langdon Down's eponym, it was Jerome Lejeune and his colleagues who actually figured out what causes the condition. In 1958, the French geneticist discovered that an extra copy of the twenty-first chromosome was the genetic troublemaker. (In 2009, fifty years after the paper that announced the discovery, the French physician Marthe Gautier, the paper's second author, published a first-person account in *Human Genetics* claiming that it was actually she, and not Lejeune, who initially observed the extra chromosome. That did not stop him from taking primary credit, she wrote. "I suspected political manoeuvring, and I was not wrong." Lejeune's supporters reject Gautier's assertions.)

Down syndrome advocates are generally appalled by the idea that a woman would terminate her pregnancy because of a diagnosis, and they worry that earlier knowledge of a Down syndrome pregnancy will make such abortions more common. While many are pleased with the brochures discussed in the previous chapter featuring cute babies with the condition and basic information about what to expect, they still sense that a stigma surrounds people with Down syndrome. To combat that, MDSC already had planned a public awareness campaign, with radio PSAs, a new website, and TV ads, called *Your Next Star*, that is aimed at promoting "the power of people with Down syndrome" in the workplace. The objective is twofold: to educate employers that people with Down syndrome can be hardworking, reliable employees, and to reassure families who receive a diagnosis of Down syndrome that their babies can grow up to be productive members of society. If moms and dads worry less about their child's future, if they feel encouraged that a meaningful life is possible for their child, perhaps it will lead to fewer abortions.

Ongoing research on how to mitigate the symptoms of Down syndrome is adding urgency to their message. These developments are both thrilling and concerning: on the one hand, advocates believe it's good that there are more options for treating Down syndrome, both because it can benefit people with the condition and because fewer parents will feel the need to abort; on the other hand, some worry it's bad because treating the condition suggests that people with an extra chromosome aren't inherently valuable as they are.

At UMass, Jeanne Lawrence is continuing to investigate how to quiet the extra chromosome that characterizes Down syndrome, studying mice in her lab to do so. An hour away at Tufts University School of Medicine in Boston, Diana Bianchi has tested various drugs on mice with a form of Down syndrome, to try to figure out how to improve symptoms in utero. And across town, at Mass General, Skotko is enrolling participants in a study that investigates whether medication can improve cognitive function in people with Down syndrome. This research and other studies are intensifying a vexed debate within the Down syndrome community. If it's scientifically feasible, is having an extra chromosome something that should be fixed?

While some parents are sharply opposed, many are confused or ambivalent about "changing" their child. But not Carolyn Bristor Hintlian. She sees no contradiction whatsoever. If a mother's job is to set her child up for success, why wouldn't she do everything possible to make that happen?

Hintlian was thirty-two when her oldest child, James, was born. She wasn't old enough for her doctor to have recommended an amniocentesis; at the time, it was advised for women thirty-five and older.

This was 1995, and recent blood tests that can sniff out Down syndrome early in pregnancy were far in the future. So James's diagnosis after he was born was a shock. Hintlian recalls the pediatrician, whom she had thoughtfully chosen while she was pregnant, entering her hospital room hours after James was delivered. It was late in the day on a Friday, and he clutched his coat in his hand, telegraphing that this was the last place in the world he wanted to be. The doctor didn't pause to sit down as he told Hintlian and her husband that he suspected from their baby's almond-shaped, upward-slanting eyes that he had Down syndrome. "We went from being on top of the world, with these perfect lives and growing careers and a baby to cap it all off, and then, suddenly, it was different," says Hintlian, who lives in Winchester, Massachusetts, outside Boston. "We were afraid." For the rest of their hospital stay, the nurses interacted with the family as little as possible. It was as if they didn't know what to say.

When James was several weeks old, a geneticist examined him.

"What will he be able to do?" Hintlian asked the doctor.

"He will be able to make a peanut butter sandwich," she responded.

Twenty years later, Hintlian still remembers how much it hurt to hear the doctor's flip assessment. "It was devastating," says Hintlian. But over the years, it morphed into a family joke. James loves peanut butter sandwiches, yet it's only recently that he himself has spread peanut butter on slices of bread and assembled them into lunch. "Of course he could make a peanut butter sandwich, but only in the last year has he actually done it because he manipulates people into doing it for him," says Hintlian. "We joke that he is way too smart to make a peanut butter sandwich."

Hintlian was able to quit her job as a food scientist to stay home with James. That hadn't been the plan, but when she learned that he had Down syndrome, she decided that no one would do as good a job stimulating and working with James—therapy is considered paramount to promote development in babies with Down syndrome—as his own mother. It became her full-time job. "If he was sleeping, I felt like I needed to be learning," she says.

James's speech as a young adult can be difficult to understand, and Hintlian feels certain that he's more intellectually delayed than the average person with Down syndrome. She says this by way of direct comparison: in the two decades since James came into their lives, she's been around lots of people with trisomy 21. He doesn't read for pleasure, for example, though he scans his mother's calendar and can pick out the word "chicken"—a favorite food—on a menu.

Throughout James's life, Hintlian has kept updated on Down syndrome research, so when she heard about a study that Skotko was helping run for a drug that may improve memory, speech, and learning in adults ages eighteen to forty-five with Down syndrome, she felt buzzy with excitement. "I assumed there would be a stampede, so I emailed. I called. I wanted him in this study," says Hintlian. She didn't ask James if he wanted to participate, because the idea of being part of an experiment felt to Hintlian too abstract for him to grasp. She rationalized her decision as follows: "In his life he has been to enough doctors—what's one more to him?" James enrolled close to the beginning of 2014.

As it turns out, there was no stampede. People were scared about potential side effects and wanted others to step forward first and be the guinea pigs. Just six families signed up at Mass General. The Hintlians were the third.

It is hard for Hintlian to envision James living independently, though that's what she wants for him—and for her and her husband. What mother doesn't?

So it was with no trepidation that she enrolled James in the study for a drug manufactured by Transition Therapeutics. She knew that Skotko was highly regarded by the Down syndrome community and figured that if there were major side effects, he wouldn't be trialing the medication. If the drug was going to work, she wanted it to work on James. And if it wouldn't have an effect? Well, he'd be no worse off.

There was one small complication: James is not exactly the most cooperative patient in the world. Hintlian was so eager for James to participate in the study that she had neglected to mention to the study administrators that her son, in nearly two decades of life, had never swallowed a pill or willingly acquiesced to having his blood drawn. Like others with Down syndrome, James has sensory issues and doesn't like being touched. Yet this particular study involved taking a pill every day and giving a blood sample at each clinic visit. "It was like a rodeo when James first came in," says Mary Ellen McDonough, who was the research coordinator for the drug trials at Mass General. "The blood draw was a couple hours of hell. I have been a pediatric nurse for forty-two years and have never encountered anything like that." She resorted to Skyping with James's beloved teacher, Mr. Carr, who calmed James down until McDonough successfully drew blood. "I was so happy when I saw blood that I was like, 'Oh, dear Jesus, thank you,'" says

McDonough. At other times, she'd Skype in the family cat, Clara, or James's siblings, who are two and four years younger than him. (They don't have Down syndrome; the condition is rarely inherited.) "James, you are so brave," they'd praise him over the computer screen. "You can do it!"

People with Down syndrome tend to have tongues that are larger than average and often protrude. It makes tossing back a pill tough. But McDonough worked with James to overcome his gag reflex, and ultimately he was able to swallow the medication. For James's mother, it was the only possible outcome. "Come hell or high water, that kid was taking the pill and getting his blood drawn," she says. "I don't want to imply that I'm a monster mother. It was just that this had suddenly become the time to master these life skills, and so yes, come hell or high water he was going to master them. That was my attitude."

The best studies are double-blinded and placebo-controlled, which means that participants are divided into two groups—one that takes the medication and one that receives a placebo. Neither the groups nor the researchers know who is getting actual medication. If they knew they were taking the drug versus an inert pill, scientists and subjects—or in this case, their families and teachers—might be swayed into thinking that it was working.

During the course of the study, James's teachers were confident that they saw a difference in James. Hintlian and her husband, Jamie, were less sure. But after the study wrapped up, they grew more certain. His thought processes seemed sharper, his word choices more targeted. He was weaving words into conversation that he wouldn't have used before and speaking more words than he usually did. He did better on a math test and appeared more aware of what was going on around him. "It was like he gained a few IQ points," says his mom.

As it turns out, Hintlian's observation appeared to be more than mere coincidence. Once the study ended, McDonough was able to tell James's family that he had indeed received the active trial drug.

In fact, all six study participants at Mass General received the drug, although no one knew that until their participation was completed. Four of the families had been convinced that their participant was taking the drug; the others weren't as sure. Yet most of the parents reported a decrease in disruptive behavior before they knew their children were on the drug; their children seemed more alert. "There was nothing that changed in James," says Hintlian. "It was just letting out what was in him. He could communicate better, be understood better, talk about his feelings. His personality was exactly the same."

Says McDonough: "The first [six] families are such pioneers. People with Down syndrome have not had access to clinical trials and it is long overdue."

The research studies are the culmination of years of work in laboratories, with mice and human cells. The Transition study was the first significant trial of scyllo-inositol, which was thought to improve working memory and perhaps stave off Alzheimer's disease; Skotko was also involved in testing a drug, manufactured by F. Hoffmann–La Roche, for kids as young as six who have Down syndrome.

Interest in finding ways to treat the intellectual delays that characterize Down syndrome is picking up. At Case Western Reserve University, the neuroscientist Alberto Costa, who has a daughter with Down syndrome, is pursuing a different drug that he believes shows promise. Another drug is being studied in Australia. And in Dallas, at the University of Texas Southwestern Medical Center, doctors are exploring whether Prozac administered to women pregnant with fetuses who have Down syndrome can boost the babies' brainpower.

We are asking parents who have kids with Down syndrome to consider testing out all sorts of medications that could make them more alert, better able to remember some things and process others—in some ways, make them smarter. If we were offering such improvements to others, it would inevitably raise questions about "designer babies"—or children or adults, for that matter. We'd be debating the propriety of enhancing intelligence. But upon closer look, is this really any different from what parents everywhere do to maximize their kids' potential and expand the opportunities available to them? My son picked up a diminutive violin and started Suzuki lessons at age four and a half. Yes, he initiated it after sitting rapt as symphony members played a Chanukah concert while he was in preschool. But I supported his interest, and I couldn't help but recall that numerous studies have linked music lessons with improved math skills. Collectively, we plunk millions of dollars into enrichment in sport and dance, art and chess. We are trying to stretch our children's horizons, to give them every opportunity. Don't these drugs do that for children with Down syndrome? Aren't they essentially an opportunity to level the playing field?

"It is a philosophical decision as much as it is a scientific one," says Skotko. "For so long, we've been telling parents that we are standing side by side with them, celebrating and accepting their sons and daughters for who they inherently are. Now we are coming to them to say we are still celebrating, but we are also offering an opportunity to maximize their life potential. Some families can easily reconcile those two concepts. They can still celebrate while offering an opportunity to maximize potential. There's no dichotomy. That's Carolyn. But other families wonder if they would inherently be changing the fundamental nature of the son or daughter with an extra chromosome who they love."

Skotko has talked about the research studies with his sister, Kristin, who is thirty-six and has Down syndrome, and with their parents. His family lives in Ohio, and it would be hard to come to Boston, where Skotko oversees the research, on a regular basis. But they are "open and excited about the promise," says Skotko, and would enroll if a trial opened up closer to them.

One drug he helped test, the Transition therapy, may lead to a better understanding of how Down syndrome relates to Alzheimer's disease. Most scientists suspect that Alzheimer's is caused in part by accumulation of beta amyloid plaques, which are encoded by the *APP* gene—a gene that is located on chromosome 21, the same chromosome that is repeated in triplicate in Down syndrome. People with Down syndrome develop Alzheimer's more frequently than the general population. From the moment they're born, people with Down syndrome overproduce the amyloid plaques. The Transition drug may fend off accumulation of the plaques. Perhaps, by preventing them from clumping together, it will be possible to stave off Alzheimer's-related dementia. But there's more troubleshooting to be done, considering that only half of people with Down syndrome start to develop Alzheimer's by the time they're about fifty years old. So the connection between the two conditions, while compelling, doesn't fully explain why nearly half do not develop the symptoms of dementia.

The Roche drug that Skotko was testing in children was thought to decrease the impact of the neurotransmitter GABA. People with Down syndrome churn out too much GABA, and the overproduction is thought to inhibit memory. If a student with Down syndrome were listening to a story, he might not remember all the twists and turns that another student without the condition would retain. "GABA makes you forget details, and too much makes you forget really important details, which makes it hard to learn," says Skotko.

In the summer of 2016, Roche discontinued the trial after concluding that the medication, basmisanil, had not resulted in any "significant difference" in cognition or function in adolescents and adults with Down syndrome enrolled in a similar study. "It's a bummer, for sure," Skotko says. "But the trial was conducted well, and the data were sound. So, in the end, the drug just didn't do what we had hoped it might be able to do, which is why these trials are conducted. I told the Down syndrome community that, disappointment aside, they paved the way for many more trials to come. Families demonstrated that our community is interested, eager, and willing to participate in clinical trials."

Skotko recruits research participants at presentations to local

Down syndrome groups and at national conferences. After a slow start, interest in participating in research has picked up, especially since the reports of Jeanne Lawrence's success in shutting off the extra chromosome in a cell in a petri dish. "People were calling Mary Ellen, saying, 'I want that medication.' We had to explain that [Lawrence's] work is still in the lab stages," says Skotko.

❐

Because Down syndrome involves an entire extra chromosome, it's long been dismissed by the research community as too complicated a genetic puzzle to solve. It's challenging enough to figure out how to treat a condition caused by a change in just one of the 19,000 or so genes that reside on our forty-six chromosomes. But factor in a surplus chromosome that contains hundreds of genes and it can seem overwhelming. That is why the recent surge in interest and subsequent approach to drug development is so significant. Rather than focusing on the whole chromosome, pharmaceutical companies are wondering whether small changes in the chromosome's functioning can result in big impacts. "Science has been somewhat slow to try to disentangle the mystery of Down syndrome," says Skotko. "But now we're on a roll. We have parent groups and researchers and funders all trying to shine the flashlight on this extra chromosome."

Melanie Perkins McLaughlin, a documentary filmmaker from outside Boston, was forty when a test for Down syndrome came back positive during her pregnancy with her third child, Gracie. "I am constantly telling Gracie that I love her Down syndrome," says McLaughlin. "I tell her she rocks that chromosome. What happens if she is thirteen and there is a pill and we can get rid of her Down syndrome? What would she say? She would say, 'I thought you said you loved my Down syndrome.'"

McLaughlin's struggle to reconcile her feelings about Down syndrome began with Gracie's prenatal diagnosis. Twenty weeks pregnant, McLaughlin had a decision to make, and she had to make it fast. After 24 weeks, Massachusetts law allows abortion only to save a mother's life. As part of her decision-making process, she connected with a mentoring program called First Call, which is offered in many states, and met with two families. Each had a five-year-old with Down syndrome. One child had fun playing hide-and-seek with McLaughlin's own kids, whom she'd brought along. The other child was nonverbal. Both sets of parents told McLaughlin that they loved their kids as is; even if erasing that extra twenty-first chromosome from each and every cell were possible, they wouldn't do it. "I figured they were saying that," she recalls, "only because they didn't have a choice."

Gracie was born on December 26, 2007. She was whisked away to the neonatal intensive care unit because of a heart defect related to Down syndrome. McLaughlin couldn't get up the nerve to lay eyes on her for nearly eight hours. "What kind of mother gives birth to a newborn baby and doesn't see her?" says McLaughlin. "I was so afraid of what she was going to look like. Finally, I went down in the middle of the night. She looked huge and healthy, just like my other babies. I was shocked."

It was 4:00 a.m. when she turned to the NICU nurse, marveling, "She's so beautiful."

"You're gonna keep her, huh?" responded the nurse.

"Lady," thought McLaughlin to herself, "you have no idea."

My first glimpse of Gracie was unscripted. She was in her room, lip-synching to Lorde's pop hit "Royals." "Let me be your ruler," she is crooning, orange headphones on, orange T-shirt on, purple glasses completing her rock-star look. She has straight hair the color of amber honey, held back with a barrette. She is seven, the same age as my youngest daughter.

Gracie is in first grade in a general education class. Monthly, on Saturdays, she participates in literacy/numeracy instruction for kids with Down syndrome. And at home, in a stunning Victorian in Medford, Massachusetts, that her parents are restoring to its former glory, Gracie has the equivalent of her own school set up in the living room. There's a small desk with a tag that has her name written on it, just like in a real classroom, and a cabinet with drawers labeled "Sight Words," "Play Dough," and "Letters." Kids with Down syndrome need plenty of educational support, so McLaughlin does what is referred to as "pre-teaching" and "re-teaching"— introducing Gracie to concepts she'll encounter in school and then reviewing those same concepts after they're taught in the classroom. On this sunny spring afternoon, Gracie is practicing her letters in a laminated workbook. "That's not 'D,' Gracie," says her mother. "Erase that and try again, please. Follow the arrow down and around."

"I can't erase it," says Gracie.

"Yes, you can," says McLaughlin. "Try again."

As a parent of a child with an extra chromosome, McLaughlin has become a mini-expert in many areas: part teacher, part doctor, part community inclusion advocate. It's a journey she never anticipated. In college, one of her roommates had majored in special education. "I remember saying with a look of disgust: Why would you want to do that?" says McLaughlin.

McLaughlin is a study in gentle and tough, working class and erudite. She's got her dark hair pulled back in a loose bun and has paired a Boston Red Sox shirt with antique diamond drop earrings. As chairwoman of the special ed advisory committee in Medford, she's fearless about fighting for what she believes Gracie and kids like her need to succeed. When Gracie was a preschooler, McLaughlin participated in litigation fighting for integrated preschool. Research shows that if you include kids with Down syndrome with typically developing kids, their outcomes are better. Put kids with Down syndrome with other kids who have speech problems, and they continue to have speech problems. Place them with kids who don't have speech problems, and they are more likely to learn to talk like the typically developing children.

When McLaughlin was waging her war for inclusion, she took Gracie to City Hall to lobby the mayor. He wasn't there, but his secretary gave Gracie a flag, as she does with every child who visits. McLaughlin snapped a picture of Gracie holding it aloft outside the building. "It's like civil rights," says McLaughlin. "We tell people all the time that it's no different. They were segregated and we want integrated. Separate is not equal."

As McLaughlin waxes passionate about her daughter's education, Gracie is grooving to Macklemore's "Thrift Shop." Then she switches to Taylor Swift's "Blank Space." "That's my favorite," she tells me.

"It can be hard to teach Gracie," says McLaughlin, "but it's incredibly rewarding." It can also be mortifying, because kids with Down syndrome may not have the social filters or appreciation of nuance that other kids do. When I visited, Gracie was experimenting with swearing, which prompted McLaughlin to recall the first time that Gracie got her teeth cleaned. The dentist tried to put her at ease, saying: "You're so cute. Can I look in your mouth?"

"No, get away, you fucking bitch," replied Gracie.

One of McLaughlin's friends, upon hearing the story, remarked that Gracie simply says what everyone else is thinking.

She was similarly prone to cursing in tap dance class, where the teacher had warned her that she could not return if she continued to talk trash. McLaughlin believes this behavior is not intentional—she thinks that Gracie has difficulty understanding why some words are okay to use in public while others are not.

There's no question that raising a child with a genetic condition is more complicated, in spades, than raising a child without. One of the recurring anxieties revolves around who will be responsible for looking after that child once she becomes an adult and aging parents succumb to their own medical conditions. Some worry about the impact that a mentally disabled child will have on other family members. McLaughlin doesn't sugarcoat the challenges when she opens her home to families who've received a prenatal diagnosis. She volunteers as part of the same First Call program that helped her when she was pregnant with Gracie and didn't know what to do.

Couples come over, sometimes with their older children in tow. They are curious, as was I, about what life is like with a child who has an extra chromosome. McLaughlin is happy to show them, with the caveat that the glimpse they will witness is one child in one family and not in any way indicative of Life with Down Syndrome, if there even is such a universal portrait. "I always tell them that it's judgment-free, that I've been there and that we changed our mind every half hour," says McLaughlin. "The most important thing is to be well-informed."

McLaughlin has no regrets about having continued her pregnancy. "All I know is I am way more compassionate now. I think, before, I saw the world in black and white and now I see in color, in a dimension that I didn't see before. I excluded people with disabilities my whole life—I didn't talk to them, didn't look at them. If I was in line with someone in a wheelchair at a grocery store, I would look away. If a person with Down syndrome was bagging my groceries, I wouldn't talk to them. Either Gracie stripped me down to the core of who I really was all along, or she changed me."

Gracie's siblings—her brother, Aiden, twelve, and sister, Ryleigh, fourteen—are used to standing up for their little sis. Aiden, in particular, is concerned about protecting her emotionally and physically. When he was in preschool and Gracie was a baby, he seemed anxious about Gracie, and a consultant called in to observe him reported that he was worried that she would die. It's not hard to see why he felt that way; over the past seven years, she's had eight surgeries, on her heart, her hips, her eyes, her ears. To this day, when the family heads to the beach, Aiden worries that the tide will sweep Gracie away. "He's a sweet big brother," says his mom, "but it's definitely affected him in terms of anxiety, understandably so."

During my visit, Gracie's family invited me to share a meal at a local chophouse, all of us seated at a big round table. A mortified Aiden tried to disappear into his fleece when his mom sent back his Shirley Temple and requested one with more grenadine. Ryleigh bubbled with excitement about the school play, *Beauty and the Beast*, and her schoolgirl crush on a fellow actor friend with ten thousand Instagram followers. Their dad, a stonemason, sat quietly amused, listening to the chatter. Seated next to me, Gracie munched on rolls and played with her homemade flash cards, offering her doll "Big Baby" slurps of orange juice.

As Aiden emerged from his jacket and sipped his replacement mocktail, talk turned to Gracie. "Maybe she'll be sad that she doesn't learn like the other kids," he said.

"You think that will make her sad?" said McLaughlin. "Gracie, do you have Down syndrome?"

"Yes," replied Gracie, as if on cue. She jauntily delivered her stock line: "And I rock that chrom-some."

"So what, Aiden," continued McLaughlin, "if there are people who don't treat her the same? Those people don't have to be part of our lives."

He stayed silent while Ryleigh piped up on the subject of Gracie and social media. "When friends see my Instagram photos and they say, 'Doesn't she have Down syndrome?' I say, 'What does that have to do with what I'm talking about?'" She punctuated her words with an acrobatic eye roll.

Several years ago, Brian Skotko published in the *American Journal of Medical Genetics Part A* what were then some of the only surveys of how people with Down syndrome and their relatives feel about the condition. More than two thousand parents (and nearly three hundred people with Down syndrome) participated in the survey. Five percent of the parents said their child embarrassed them in general, and 4 percent said they regretted having a child with Down syndrome. Can you imagine the emotional steeliness, the depth of despair, required to check the box for regret? Still, the overwhelming majority are happy with their child, as are the siblings: again, just 4 percent said they wished they could swap their sibling with Down syndrome for one without that extra chromosome. And 88 percent said they believed that having a brother or sister with Down syndrome had made them a better person.

□

When Jerome Lejeune identified the chromosomal basis for Down syndrome, it paved the way for other scientists to develop methods of prenatal diagnosis. This horrified Lejeune. A devout Catholic, he was an ardent supporter of people with Down syndrome and just as ardent an opponent of abortion, particularly as an outcome of prenatal testing. Someday, he felt certain, a cure would be found. Lejeune, who died in 1994, was a quotable fellow. "We will beat this disease," he was quoted as saying. "It's inconceivable that we won't. It will take much less intellectual effort than sending a man to the Moon."

Diana Bianchi is trying to fulfill Lejeune's prophecy. It would be a stretch to say that Bianchi, until recently a Tufts geneticist and neonatologist, has hit upon a cure, but she is working with cute furry mice, lots of them, to come up with a way to alter the course of Down syndrome prenatally. Bianchi, who took over in late 2016 as director of the National Institute of Child Health and Human Development, has been derided by some in the Down syndrome community for her role in helping to introduce non-invasive prenatal screening. They believe that facilitating a Down syndrome diagnosis is tantamount to enabling abortion. But Bianchi maintains that identifying Down syndrome early in pregnancy motivates researchers to search for a prenatal therapy.

Mark Bradford has a son with Down syndrome and is the president of the Jerome Lejeune Foundation USA, the American arm of a Paris-based group named for the French physician. "There are some so-called pro-life advocates in the Down syndrome community who consider [Bianchi] a villain because of her involvement in the development of NIPT," Bradford says, referring to noninvasive prenatal testing, another term for noninvasive prenatal screening. "I think one day she will prove to have been a heroine in that her advancement of NIPT will be the gateway to early therapy and will save countless lives of babies who have been prenatally diagnosed with Down syndrome. She is very unjustly and harshly criticized for her work by people who can't see past the threat of prenatal diagnosis to its future potential benefit."

Increasingly, there is optimism that prenatal detection may positively impact the course of a pregnancy. Some cases have already fulfilled that promise: in spina bifida, for example, prenatal detection paired with delicate in utero surgery can stitch up the neural tube defect and make the difference between a child being confined to a wheelchair and being able to walk on his own. Likewise, fetal echocardiograms can reveal heart problems that can be repaired prenatally. Evolving gene-editing technologies such as CRISPR-Cas9 may one day be able to correct problems before birth.

Continued progress in developing therapies for Down syndrome raises an important question: Is there a point at which better treatment options will make abortion a less common choice? Assuming that an effective treatment emerges for Down syndrome—long considered an exercise in futility because every single cell in an affected person's body contains an extra copy of chromosome 21—will many people still want to avert a Down syndrome pregnancy?

When I asked Bianchi if her ultimate intention is to decrease the rate of abortions for Down syndrome, she hedged. "Our goal is to hopefully improve neurocognition and, in doing so, provide expectant couples with a message of hope," she says. "What people decide to do with that information is their business."

Bianchi never met Lejeune, but she keeps a framed handwritten reply to a request she sent as a college student in 1973, asking—in very good French—for a summer internship in his Paris laboratory. The famed physician began his reply: "Mademoiselle, all my greetings for your French, which is excellent." He went on to express regret at not having room for her in his lab at that time. Bianchi keeps everything, and she kept this letter, tucked away in a box, for years. "One day a couple years ago, I found it," she says. "I thought, 'This is a sign.'"

Bianchi has curly blond hair that's perfectly waved and parted in the middle. Her nails are painted bubblegum pink; she is wearing a soft teal turtleneck and taupe blazer. She's sixty, an age when others are slowing down their careers. But Bianchi is revving up.

I am curious about her mice, so I have asked Bianchi if I can check them out. It takes several weeks to secure the necessary approvals from Tufts's Division of Laboratory Animal Medicine, which is concerned about outsiders transmitting infections to the mice; Bianchi tells me I am lucky, for I am the only journalist who has ever seen these research rodents. They are cosseted critters, whose keepers demand that visitors don full protective gear to protect the mice from the detritus of the outside world. Before we step into her lab, we slip on gauzy elasticized poufs covering our curls, long-sleeve gowns, booties, and gloves. The ringmaster of this rodent circus is Bianchi's postdoctoral student, Faycal Guedj, who puts the mice through their paces.

As a journalist who writes about science, I am aware that all sorts of creatures are bred for genetics research, special-ordered with a particular genetic mutation. I encountered a tiny worm of the species *C. elegans* at the Children's Hospital of Philadelphia that has the same genetic alteration as Juliet, a stunning ebony-haired, ebony-lashed beauty who can't walk or talk but nestled into me as I read to her on her parents' couch last year. I learned about a frog named Bea with the same TGF-beta mutation as Beatrice, whose muscles are weak, her eyes widely spaced, her fingers bent. And in this tiny lab no bigger than a closet, I met dozens of baby mice—pups—christened Ts1Cje (after Charles J. Epstein, a medical geneticist—and target of the Unabomber—in whose lab this particular strain was created).

These scrappy rodents have the mouse equivalent of Down syndrome. Adult males with Ts1Cje are mated with female mice with typical chromosomes because Bianchi wants a normal uterine environment to rule out any secondary effects of maternal chromosome abnormalities. When the male mice breed with the female mice, statistically half of the pups will have a model form of Down syndrome, which ensures a steady supply of research subjects.

Bianchi is examining different signaling pathways in the mouse brains by testing the mice from day three to day twenty-one after birth and again in adulthood. Some of their mothers were fed a common drug or nutritional supplement in their chow while pregnant. The ultimate goal is to home in on a drug that can be administered orally to pregnant women whose fetuses have been diagnosed with Down syndrome.

The drug would cross the placenta and, ideally, boost cognition and reduce oxidative stress. Babies with Down syndrome display much more oxidative stress than those without. Oxidative stress occurs when so many free radicals—highly reactive molecules in search of their missing electron—accumulate that their toxicity can't be effectively counteracted. Too much oxidative stress kills brain cells and results in babies with Down syndrome being born with smaller brains than unaffected newborns. "Our hypothesis is that at the time when the brain is undergoing significant development, the abnormal biochemical environment knocks out a group of neural stem cells," says Bianchi, who explains that from about fifteen weeks of pregnancy, the brain starts to grow a bit more slowly in fetuses with Down syndrome. That's why she believes it's essential to intervene prenatally to stimulate a meaningful spike in neuron generation. But before any intervention can be tested on pregnant mothers and their babies, it must first prove safe and effective in an animal proxy.

On the day I visit Bianchi's lab, we are examining seven-day-old mice. Each mouse is tattooed with a number for the purpose of identification. Researchers don't know if the mouse has the equivalent of Down syndrome until the experiment concludes and genetic testing is done on tissue from that mouse. If they did know, they might be less than objective, automatically assuming that a mouse with Down syndrome would be slower than its unaffected counterpart.

Guedj scoops up a pup, Mouse #1, and weighs him; he is five grams, as heavy as a nickel. Guedj places the pup on his back to see how long it takes him to flip onto his belly. This fellow is speedy and rights himself in less than four seconds. "See how quick it was," marvels Bianchi. Mouse #1 does a masterful job of not falling off a blue rectangular cube about seven centimeters high; the cube simulates a cliff. A mouse fully aware of its surroundings will not want to tumble off. Then it's Mouse #2's turn. He weighs just two grams and at 53 millimeters is only three-quarters as long as Mouse #1. Bianchi says the pup could be growth-restricted—problems with the placenta could in theory render mice with standard chromosomes on the small side—but she suspects he has Down syndrome because those mice tend to be tinier. That suspicion appears correct as Mouse #2 runs through his routine. Seventeen seconds tick by as he flips from back to belly, struggling to free his right foot, which is trapped beneath his body. He blithely scoots off the cliff, without the cautious sensing displayed by his predecessor. Guedj catches him in a gloved hand.

Bianchi hopes that by treating oxidative stress in the womb, she can prevent brain cells from shriveling away. Who would argue with that ambition? Wouldn't any parent want to protect her unborn child's brain cells? It's important to note that the prenatal therapies that Bianchi is studying, no matter how successful, wouldn't change physical appearance.

"Plenty of people don't agree and think that their children

with Down syndrome are perfect the way they are," she says. "But there are also plenty of people who, if given the choice, would want to attempt to treat their children. Our job is to give people the options."

One option—the drug trials that Brian Skotko and others are administering—is too little, too late, in Bianchi's opinion. Why wait until childhood or later to see if drugs do any good? "From our perspective, waiting to test older children is waiting too long. We have made the case that things are already going awry early in development. If we could stop or minimize that, maybe everything else would fall into place."

Of the three compounds she's studied so far, one had a negative effect on cell division, slowing the process and resulting in fewer cells growing. The other two have shown some therapeutic effects; one of the compounds improved coordination and shortened the amount of time it took for mice to roll over. With some areas of development apparently unaffected by prenatal therapy, Bianchi's work goes on. She's got a new lab at the National Human Genome Research Institute, where she'll expand her research into the prenatal treatment of Down syndrome with a new raft of rodents. "From what I've seen so far, there is no magic bullet yet where you take the drug and everything is cured," she says.

❐

Melanie McLaughlin doesn't hesitate to rely upon modern medicine to address the physical concerns that can attend Down syndrome: Gracie's hips dislocate because she has loose joints, which are common to kids with trisomy 21, so she has had two grueling surgeries to forestall trouble walking later in life. But McLaughlin draws a distinction between tending to Gracie's body and tending to her mind. She is concerned that using drugs to change how Gracie thinks is not too different from eugenics—in this case, trying to build better babies.

Jeanne Lawrence, who succeeded in silencing the surplus

copy of chromosome 21 in the laboratory, has a different view. Lawrence is continuing her work with mouse models with Down syndrome, trying to neutralize the genes on their extra chromosome so that they won't be expressed. (It's kind of like popping a balloon. A burst balloon is still technically a balloon, but it can't serve its intended purpose.) Lawrence is experimenting with achieving this at all stages of development, from conception to pregnancy to after birth. When she's received worried comments from parents concerned about "changing" their children, this is how she's responded: "I said I couldn't change your child if I wanted to. I explained that if you were going to treat a fetus with Down syndrome, it would already have Down syndrome by the time of treatment and some of the properties would be present no matter what."

Of course, it's second nature to want the best for your children, and McLaughlin is not immune to this desire. But she wonders what the "best" means for Gracie—and for her two older children. "What I really want for my children is for them to be happy, to find love, to be fulfilled." But when we talk about achievement, we don't only reference happiness or love. So what does achieving the best in life imply? Does it mean that you are the smartest person in the room? The wealthiest? The one with the most impressive job title? McLaughlin had hoped for unconditional support when she decided to continue her pregnancy with Gracie, but she found that the people she thought of as her most liberal, enlightened friends were the least accepting. At the time, she was working as a documentarian at WGBH/PBS with no shortage of Ivy Leaguers. They couldn't understand why she'd want a child with intellectual disabilities.

There's no doubt that society places a huge premium on intelligence, but for people with Down syndrome, a little more knowledge or awareness is not necessarily always helpful, McLaughlin has concluded. When she talks to parents of older children with Down syndrome, she hears that those children who are more intellectually advanced can struggle more. One example jumps to mind, of a friend, an older boy, who understands that he has Down syndrome while most other people don't have it, and that it's not the most marketable hook with which to attract a girlfriend. "He doesn't want to date girls with Down syndrome, but girls without Down syndrome don't want to date him," says McLaughlin. "How much is ignorance bliss?"

5

What Do Parents
Want to Know?

Grappling with Variants
of Uncertain Significance

Pregnancy is typically a time of great expectations. But right from the beginning, little Daniel's future was suspect. When his mother, Maya Hewitt,* a professor of developmental psychology, got pregnant in 2011 and went for her initial ultrasound, Daniel was deemed an imminent miscarriage. No heartbeat was detected. But Daniel rallied, growing from a tenuous embryo, a few hundred cells strong, into a lover of trains and planes and a loather of naps.

After that initial scare, ultrasound had revealed that Daniel's umbilical cord had just two blood vessels instead of the standard two arteries and one vein. That prompted perinatal specialists to continue to closely monitor Maya's pregnancy. A two-vessel cord, also known as a single umbilical artery because the cord consists of just one artery that transfers waste from the baby to the placenta, can be associated with birth defects of the heart and central nervous system and with chromosomal problems. Maya had many more ultrasounds than the typical pregnant woman, and everything seemed to be okay.

*The names of Maya and the members of her family have been changed.

Daniel was born on April 18, 2012. When I met him at his family's small whitewashed, black-shuttered house in the suburbs of Philadelphia, he was two and a half, outgoing and engaged and flirtatious with the inquisitive stranger who arrived shortly before naptime. He had silky blond hair, a wardrobe befitting a pint-size lumberjack—plaid shirt and jeans—and a spirit of generosity. We built Ikea train tracks together, Daniel handing me sections to lay down, then zooming trains up and down and through the wooden loop-de-loops. While outside on his deck playing, he repeatedly offered me morsels of his granola bar and pointed at the impossibly blue sky while making airplane noises. I glanced skyward and spotted a jet soaring through the clouds. I hadn't noticed the plane until Daniel heard it. I have pretty sharp hearing; I can detect my children's footsteps on the stairs in the middle of the night and know who it is by the sound of the gait. Daniel, on the other hand, has hearing loss. It's why I had wanted to meet him and is what had propelled his family into the world of genetics.

"Until I actually heard him cry and heard his Apgars, we didn't know what was going to happen, and we were so relieved he was okay," says Maya, who is thirty-six. "Then before we left the hospital, we found out he had failed his hearing test. We don't have anyone in our family with hearing loss, so it was like, 'What do we do with this?'"

Daniel was referred to the Children's Hospital of Philadelphia (CHOP); the world-renowned children's medical center is near his home. When he was a month old, he began visiting a dizzying array of specialists, including an audiologist; an ear, nose, and throat doctor; a cardiologist; a urologist; and an ophthalmologist, because their specialties are associated with various systems that can be related to hearing loss, and to address other medical concerns.

Maya and her husband, Andrew, an elementary school art teacher, had no reason to suspect any hereditary cause for Daniel's hearing problems, so they saved the genetics clinic for last. "We didn't anticipate anything coming from it," says Maya, whose elfin face is framed by dark hair in a smooth pixie cut, her blue eyes outlined by funky brown glasses. They were in for a life-altering surprise.

In January 2013, Maya and Andrew took Daniel, then nine months old, to see Ian Krantz, a towering, craggy-faced pediatrician and clinical geneticist who runs a hearing loss clinic at CHOP, and one of his genetic counselors, Sarah Noon. Krantz noted that Daniel had a malrotated kidney that had been previously observed at Maya's OB clinic on prenatal ultrasound (it flops over a bit, but urology, which reviewed the prenatal scans, hadn't been concerned). He also had a divot at the base of his spine—the endearingly dubbed "sacral dimple"—that by itself was not alarming. Sacral dimples are not uncommon, but they can, in some instances, indicate an underlying problem. This constellation of features, combined with additional observations such as Daniel's two-vessel cord, a physical exam that noted he had a particularly small head circumference, and his hearing loss, led Krantz to suggest genetic testing, specifically a chromosomal microarray, the test that looks for missing or extra fragments of genetic material. And that's when life for the Hewitt family started to get really complicated.

Daniel's blood was drawn for the testing, and months passed with no word. That summer, Maya called CHOP several times to inquire: Where were the results? "We really want to know," she pleaded in a voice mail. Daniel was developing just fine, meeting milestones according to schedule, but the anxiety of waiting for results felt exhausting to Maya and Andrew.

In September, Sarah Noon phoned. Maya was in the grocery store parking lot. She listened to the genetic counselor, confused as Noon explained that the testing had not revealed any explanation for the congenital sensorineural hearing loss that Daniel had been diagnosed with when he was two days old, but had uncovered an incidental finding.

"What does that mean?" Maya asked.

Noon explained that sometimes, genetic testing—in particular, genome-wide tests such as the chromosomal microarray, with its ability to identify deletions or duplications involving snippets of genes, and sequencing, with its capacity to scan all those genes at once—uncovers findings that have nothing to do with what doctors are looking for. In some cases, those findings amount to a whole bunch of nothing, because the identified genetic errors haven't been associated with disease. In other instances, they reveal increased risk for disease or even actual disease that's unrelated to the primary reason for the test. The adage "Be careful what you look for" comes to mind. In Daniel's case, testing had detected a deletion on his fifth chromosome in a gene known as *TERT*, which is associated with telomeres, the end parts of chromosomes. You can think of telomeres as protective caps located at the tips of chromosomes that help preserve them—kind of like the tiny black caps that twist onto a tire's valve stem.

TERT mutations, or changes in the gene, have been associated with dyskeratosis congenita, which leads to changes in skin and hair pigmentation and an increased risk of cancer, bone marrow failure, and pulmonary fibrosis. Patients with dyskeratosis congenita tend to have shortened telomeres, which cause cells to age more quickly. A *TERT* mutation can particularly impact those cells that divide rapidly, like skin and hair and bone marrow. Because weakened telomeres make chromosomes less stable, cells may start to divide haphazardly, giving rise to cancer.

Yet none of this doomsday scenario was inexorably destined to play out. Daniel did not have a *TERT* mutation, a change in the gene; he had a deletion in which he was missing several genes. But whether he would be affected couldn't be predicted because there was scant information about his deletion in medical journals. Some *TERT* deletions and broader deletions on chromosome 5 had been associated with dyskeratosis congenita, but other deletions hadn't resulted in the condition. Missing genes sounded ominous and foreboding, but it was entirely possible that their absence wouldn't matter one whit.

Noon acknowledged to me, as I sat in her windowless office in downtown Philadelphia getting briefed on Daniel's situation, that having this sort of information isn't necessarily helpful. She noted that Maya and Andrew had been on edge from the get-go when it had appeared that Daniel, from the earliest embryonic stages, wasn't developing according to plan. "They felt they couldn't enjoy the pregnancy," Noon said with compassion. "Then she had her son and came to genetics and was hit again with an unknown result. She said if she had known about this, she may have decided to not have the testing. It's anxiety-provoking to constantly be in a state of wondering."

The entire genetics community shares the concern about the impact that unclear results have on families. Academic journals and annual meetings of professional associations such as the American Society of Human Genetics (ASHG) and the American College of Medical Genetics and Genomics (ACMG) are filled with papers and presentations acknowledging the problem, defining the issues, and offering insights on best practices.

Sometimes, coping with uncertainty can be more difficult than processing bad news. "When you have disappointing information," Noon observed, "in some ways it's easier, because at least you know rather than being left watching and waiting."

Maya's dismay at Daniel's inconclusive results highlighted for Noon an inconsistency in the way patients are informed about the potential outcomes of various tests. For exceptionally complex and newer tests such as genome sequencing, which reads the entirety of a person's DNA, or exome sequencing, which scans just the protein-coding portions, there are detailed consent forms that explain the scope of the test and the different sorts of information it may uncover. But no such protocol has typically existed for microarrays, which can also yield puzzling results but have been part of the genetic testing toolbox for years. "We have detailed consent forms for exome sequencing and we are thinking," says Noon, "why don't we have this for arrays?"

A consent process for an exome- or genome-sequencing test is not at all like the one-page HIPAA form you sign pro forma with your eyes closed at your doctor's office. It's many pages long and very detailed to help adequately prepare patients for what testing may reveal. It's intended to alert patients to precisely the sort of situation that Daniel's parents found themselves facing: a result seemingly unrelated to the reason Daniel was being tested in the first place, and one that could have serious implications for their ebullient toddler—or could end up meaning nothing at all. Recently, ASHG recommended extending the consent process to array testing.

Maya and Andrew had run smack into the gray zone of genetics: in most cases, genes aren't destiny; they're just part of a complex mix of biology, environment, and circumstance that plays a part in determining who gets sick and when. As genetic testing becomes more sophisticated and widespread, more families will come face-to-face with genetic uncertainty.

"Daniel is so young," says Maya, as Daniel plays happily alongside us, stacking blocks and chattering to himself. "He's just at the beginning of his life, but his whole life has been one dire prognostication after another, beginning in the first few weeks of pregnancy."

Maya and Andrew had assumed the testing would be targeted to what Krantz suspected might be going on with Daniel. "We didn't realize it was going to be the whole exome, everything," says Maya. (Actually, he'd had a microarray, a different test. Maya's momentary confusion over what sort of test Daniel had is typical; for someone not well-versed in genetics—most people, that is—one powerful test seems like another.)

The act of communicating the outcome of genetic tests, called the "return of results" in genetics argot, is equal parts art and science. When performing comprehensive genetic tests such as sequencing or microarrays, results amount to a data dump of information. As these tests become more commonly used, there's more information

that needs to be sifted through and interpreted. There's also a co-
lossal divide between those experts and physicians who feel it all
should be shared with the patient, no matter how inconclusive or
ambiguous it may be, and those who feel the data should first be
cherry-picked, after which only certain categories of information
should be revealed.

Results can be sorted into several categories: Are they clinically relevant, meaning they're related to a particular diagnosis or
condition? Are they medically actionable, meaning there's a treatment or even a cure? Are they totally unexpected, an "incidental
finding," like the one Maya encountered, that has nothing to do
with the impetus for the test? While incidental, or secondary, findings are often health-related, it's now quite common for genetic
tests to identify "nonpaternity"—in other words, the tests can
reveal that the guy who thinks he's Dad really isn't. This could
also extend to the discovery that a husband and wife are related
by blood—unwitting cousins, perhaps. Should doctors share that
earth-shattering news and risk tearing apart a family? And what
to do with the murkiest finding of all, a "variant of uncertain significance" (also called a "variant of unknown significance"), especially one so newly identified that researchers have not yet amassed
a body of evidence that indicates whether it may affect future
health?

To confuse the situation even more, some genes are, in the lingo
of the genetics world, "highly predictive." That means if you have
a particular gene or gene mutation that correlates with a particular
condition, you're almost certain to develop that condition. Huntington's disease is the best example. As we'll see later on in the
book, people in whose family the fatal brain disease runs vary
when it comes to wanting to know whether they've inherited the
genetic error on chromosome 4 that causes the disease. But then
there's autism, the highly complex condition that's been associated
with dozens of genetic mutations. You may have some of these genetic changes, yet it's unlikely that you have autism. In contrast to

Huntington's disease, there is no single genetic cause of autism. A generation ago, doctors didn't diagnose children with autism at the rates they do today; the latest statistics conclude that one in sixty-eight children—one in forty-two boys—falls somewhere on the autism spectrum. This spike in autism cases is largely a function of increased awareness on the part of parents and providers and partly a result of changing diagnostic criteria. But that probably cannot account for it all. The missing link, according to researchers, may be genetic. Scientists are tracking down genetic roots, with one ambitious project working to sequence the genomes of 10,000 people throughout the world who are part of families affected by autism. Some studies have observed a considerable probability—between 36 percent and as high as 95 percent—that if one identical twin has autism, the other one will too. Autism tends to run in non-twinned siblings too: if one child is affected, a subsequent child has a one in twenty chance of receiving a diagnosis, considerably greater than the risk of the general population. Meanwhile, recent research fingered a mutation that's associated with an increased risk of a subtype of autism and is also accompanied by physical traits—larger heads and prominent foreheads—raising the possibility of identifying at least some types of autism risk in utero.

There appears to be something genetic going on. But genetic mutations aren't just something you're born with and pass on; they can be spurred by changes in a person's environment. Men older than age forty, for example, stand a greater risk of having a child with autism because mutations in sperm can increase with a man's age. Women in their forties also carry an elevated risk of having a child with autism, although the reasons aren't as clear. Also uncertain is why couples with age gaps of at least a decade are more likely to have a child with autism. In fact, many of the most recently identified genetic links to autism are de novo mutations, meaning that they're random and spontaneous, not inherited. What's inherited and what's simply a change in DNA over time is of interest to parents and researchers trying to gain insight into what makes a person gay, obese, athletic, intelligent. In Connecticut, the state medical examiner asked geneticists to scrutinize Adam Lanza's DNA for clues to what compelled him to gun down the children and teachers of Sandy Hook Elementary School. They're unlikely to find anything tying Lanza's unfathomable violence to a genetic propensity for crime, which underscores how hard it can be to equate genetics with behavior. In an editorial about Lanza's DNA analysis, the journal *Nature* cautioned that "there is a dangerous tendency to oversimplify, especially in the wake of tragedy. If Lanza's DNA reveals genetic variants—as it inevitably will—people who carry similar variants could be stigmatized, even if those variants are associated only with ear shape. If Lanza has genetic variants already associated with autism or depression, people with those diseases could come under suspicion as well."

There's an inclination to overly blame or credit genetics, but genes aren't typically the last word on how we turn out, why we get sick, and with what. As it relates to child-rearing, this is reflected in the eternal nature-versus-nurture debate. Is a child a terror on the football field because he shares the genes that equipped his father to play in the NFL, or because that dad has taught him from an early age how to handle a pigskin? Sometimes it's next to impossible to separate the two. But the way we physically move through the world—our characteristics and preferences—can't always be credited to our genes. If you're obsessed with watching cat videos, it's not likely that you inherited your feline fetish. Harvard's Personal Genetics Education Project puts it like this: "Chances are that your personal disposition to cats is the result of your life's experience with pets rather than a mutation in a hypothetical cat fancier gene."

As powerful tests like sequencing get incorporated into health care—a 2016 survey of U.S. physicians found that 27 percent had recommended genome sequencing to their patients—doctors must grapple with adequately preparing patients for results that are often inherently ambiguous as well as both expected and unexpected. Is it fair and ethical to withhold some results from patients, particularly unanticipated results that are uncovered only in the process of looking for something else? If so, what sort of results should doctors return? These decisions come into sharpest focus when it comes to children, who are the subject of genetic tests in the womb and afterward if their development is delayed or they're sick with an undiagnosed disease. Not surprisingly, physicians and bioethicists agree that results that stand to affect a child in childhood should be shared—such as the unanticipated discovery of an altered gene that gives rise to a rare form of childhood colon cancer. Considering the time constraints inherent in modern medicine, *how* should that potentially game-changing information be shared—in a phone call? A letter? In person?

And what about results that indicate disease risk in adulthood? What about results that reveal risk for an as-now untreatable diagnosis, such as Alzheimer's disease? Who should make the call about what results to return?

❒

Maya Hewitt believes the parents should. In order to be able to do that, they need to be fully informed about the kinds of potential information that testing may reveal. That day at Maya's house, she described to me a sensation, an undercurrent of apprehension, that felt as if all the air from her lungs had been whooshed out. She says she doesn't ever feel fully at ease now that she knows about Daniel's gene deletion. Yet at the same time, Daniel is thriving. At his preschool parent-teacher conference, Maya beamed as Daniel's teachers reported he was well ahead of where a typical toddler should be.

"He's a normal little kid except he's missing a piece of his fifth chromosome, which we never would have known had we not done the testing. But it changes things for us, for certain. It causes us to live with this always in the background because of the nature of dyskeratosis congenita and the other issues that could unfold later. We are trying to maintain as much normalcy as possible, but it's operating in the background.

"When he reaches some new developmental milestone, I think, 'When is this going to get taken away?' When I have a mother-son moment, I enjoy it for thirty seconds and then I think, 'This could change in a few years.' That could be the case for anybody, but we know that he has something that we know could actually affect him. It's not theoretical.

"I appreciate science. We are not people who want to bury our heads in the sand. But there is a danger of knowing too much about something that they, the experts, don't know enough about. Maybe I expect too much from science that they're going to know what everything means. But our experience is that we are caught in a gray area—our report says 'of unknown clinical significance.' I wish we could stuff it back in the box and just live in ignorance. I wish I didn't know."

After Maya heard the news from Noon that fateful day outside the grocery store, she stood dumbfounded, momentarily frozen in the parking lot. Then she dialed Andrew, who was at home with Daniel, and did her best to explain the situation. Within a few days, grief overwhelmed her. "I felt really sad that the childhood we envisioned for him was potentially going to be ripped away at some point. Andrew describes it as we began to live with an anvil over our heads. It's like living underneath a shadow."

Noon had also suggested that Maya and Andrew come in for parental testing. If either parent had the same deletion, that could offer context about what it might mean for Daniel. Initially, Maya and Andrew blanched at learning about their own genetic pedigrees—they were of the mind that if it ain't broke, why fix it?—but if it could help Daniel, they'd do it. In any case, they doubted the deletion had been inherited. After all, symptoms from tooth decay to nail disintegration to bone marrow failure could start making themselves known anytime from age five onward. They were both in their thirties, and they were fine. As far as they knew.

Within a few days, they gave blood. Again, months went by with no word. (Maya says she was eventually told that the delays stemmed from staffing issues.) Maya and Andrew started to doubt themselves. A friend of theirs who has a Ph.D. in nursing advised them to try to put the testing in perspective and move on with their lives. Other friends concurred, pointing out that the angst they were feeling was overwhelming. The waiting, the watching, the not knowing hovered like a storm cloud. They decided they would do their best to put the *TERT* deletion and the genetic testing out of their minds. "We made the decision that we weren't going to follow up on it, and we just wanted to live normally," says Maya. "If it happened and disease manifested, we'd deal with it. But Daniel was healthy, and we decided, 'Let's just live.'"

That approach sounded good in theory. In reality, rather than soothe Maya's anxiety, the ongoing uncertainty only inflamed her anger. She points out that while she's not a medical professional, she is a researcher. Researchers collect evidence. They don't leave things to chance.

As April of 2014 approached and with it Daniel's second birthday, Maya still hadn't received the results from the testing that she and Andrew had undergone. She decided to write a letter to the providers at CHOP. She was precise and professional, but the contents of her letter breathed fire. In it, she unleashed her frustrations at having her quintessential storybook life—the two-career family, the loving husband, the beautiful child—upended.

> Given that some of the outcomes associated with the deletion that your office detected include various life-threatening illnesses, including some that may not arise until ages 5–15 or young adulthood as you indicated, we are beside ourselves with grief at the possibilities that you have now opened up for us and, in many ways, feel that you have

unethically robbed us of our family's peace of mind by performing tests beyond what we were anticipating.

Maya recounted her experience to me as we sat on her living room floor, helping Daniel construct bridges and byways from interlocking wooden tracks. "I was just shocked, blindsided that there was anything beyond the hearing loss, let alone something that sounded so treacherous," she tells me.

"We would have liked more explanation," says Maya. "If you're going to do such a high-powered test that can tell you a ton of information, some of which you might not be prepared for or know what to do with, there's a very important educational piece. We didn't know what we were stepping into, and we are very educated people. What about people who don't have the education and knowledge and resources to understand this?"

I found out about Daniel from the last person you'd think would want to tip me off to his parents' displeasure: Daniel's doctor, Ian Krantz. Krantz and I had spoken on the phone several times and communicated by email more times than he probably cares to recall, but we'd never met until 2014 at the annual meeting of the American Society of Human Genetics. At a cocktail party hosted by the Children's Hospital of Philadelphia and the University of Pennsylvania, we joked about the ribbing he still endures over a moody photo of him and his wife that had accompanied a *Time* magazine cover story I'd written about sequencing children's genomes. Scientists aren't generally accustomed to being posed or asked to look purposely pensive or studiously thoughtful. But he and his wife, Nancy Spinner, a fast-talking, no-nonsense scientist who ran the cytogenomics lab at CHOP at the time, rose to the occasion. The result was a dreamily lit image taken in Krantz's lab of Spinner, left hand on hip, seemingly peering into the distance, while Krantz, right hand in pocket, casts his gaze downward. Years later, their colleagues still tease them about that photo in which they look like Big Thinkers. In truth, they are.

Krantz and Spinner fell in love in the lab, researching Alagille syndrome, a hereditary childhood liver disease. They found the genetic cause. They also found each other (nor have they given up on curing Alagille). They have three kids, which draws their work into sharp focus. At CHOP, they help parents map out their kids' genetic profiles.

One of the couple's most vexing cases occurred when the laboratory inadvertently discovered that a sick baby carried a high-risk gene for a rare form of early-onset dementia. Their laboratory and clinical teams agonized about what to do, ultimately deciding not to inform the parents that their child was at risk for dementia, perhaps as early as age forty. With no current treatment or cure, telling them about a memory-sapping disease that hovers like a storm cloud in their child's future felt destructive. "We came around to the realization that we could not divulge that information," says Spinner, whose lab staff tested the infant. "One of the basic principles of medicine is to do no harm."

Around the same time, a toddler was tested in her lab and found to have the previously mentioned gene linked to a rare form of childhood colon cancer. In some cases, polyps arising from this kind of cancer have been known to develop in elementary schoolers. This time the decision was not as fraught; keeping quiet would have felt unethical. "We feel good about that one," says Spinner. "Proper screening can make a huge difference."

In the case of the child with the dementia gene, Krantz and Spinner felt they were protecting the parents from emotional distress, particularly since there is no treatment for dementia. As with dementia, there is no treatment for Daniel's *TERT* gene deletion. Should Krantz have applied the same lesson and kept quiet, thus sparing Daniel's parents the ongoing uneasiness they feel? Says Spinner: "I'm starting to think we need to back off. I'm starting to think we shouldn't provide all these variants of unknown significance. I'm starting to think we should be more careful about overcalling things." At the same time, there is a case for revealing mutations to parents—even when they are associated with diseases that do not currently have treatments. Say that a therapy, even a cure, is developed ten years from now. If parents don't know about their child's gene mutation, they won't know to pursue that therapy.

The gut-wrenching decision-making process underscores the importance of the research that Krantz and Spinner are leading to figure out how to navigate this new world. They're recruiting families with kids who've got hearing loss and autism, kids who've suffered heart attacks and have mental retardation. They're mapping their genome, then asking their parents what information they want to know—and what they don't. A multiyear grant from the National Human Genome Research Institute, part of the National Institutes of Health, has allowed them to dive into the particulars surrounding how to deliver genome-sequencing results. They are examining the ethical and psychosocial impacts of a test as powerful and comprehensive as genome sequencing, looking at the role of counseling before the test and whether it's appropriate to share findings that aren't connected to the reason why the test was done in the first place.

In her angry letter, Maya referenced Greek mythology to describe the predicament CHOP had put her family in.

> As Daniel's second birthday approaches, we see before us a vibrant and much beloved little boy who is thriving by all other accounts. The uncertain "Pandora's Box" that you have unwittingly opened up for us, of course, can never be closed and we will likely always live with some amount of heightened anxiety about what could be in his future now.

The mythology of Pandora's box is frequently invoked when it comes to genetic testing. Once upon a time, in ancient Greece, Zeus—no milquetoast, by any stretch of the imagination—decided he had to mete out punishment to two brothers, Epimetheus and Prometheus. He molded a fetching woman from clay and called upon the talents of other gods to fine-tune his creation. Athena breathed life into her, while Aphrodite bestowed great beauty upon her and Hermes imbued her with charm and cunning. Zeus named her Pandora.

Epimetheus was smitten, despite his brother's wariness. Wedding plans percolated, and with them, wedding presents. Zeus generously gave Pandora an ornate box and cautioned her never to open it. Spoiler alert: this gift was not what it seemed.

Alas, Pandora's curiosity got the better of her. Why would Zeus give her something so precious if she couldn't even see it? If the contents of the box were truly so very valuable, surely she should at least take a peek, right?

Alone, Pandora slipped the key that had accompanied the gift into the lock and unlatched the box. But she didn't have the nerve to follow through, so she locked the box again. She repeated this process three times until she could no longer resist temptation. What could possibly be nestled inside?

What she found bore no resemblance to what she'd anticipated. Instead of riches or bedazzling jewels, she was accosted by Sadness and Death, Envy and Disease, Poverty and Hatred, buzzing about and stinging her with no mercy. She slammed the box shut, to no avail.

Misery had been unleashed upon the world.

The story of Pandora's box calls to mind Eve's insatiable curiosity about the apple in the Garden of Eden. At its heart, Pandora's box and the tale of what transpired in idyllic Eden are all about succumbing to curiosity, only to find yourself grappling with a situation over which you have lost control.

When it comes to genetic knowledge, the moral of Pandora's box—don't be overly curious—may be too simplistic. While learning about previously unknown risk factors or disease can feel overwhelming, that knowledge may be useful and change the course of a person's health care. That was the case with my daughter.

Shortly after I began writing this book, I discovered that sometimes, so-called incidental findings aren't entirely incidental. Nor can they be neatly characterized as tampering with Pandora's box.

Call it mother's intuition: even though my middle child, Shira, at nine, had no classic gastrointestinal symptoms, I had asked that she be tested for celiac disease when her blood was drawn to try to figure out why she'd been running a fever for a week. Fever is not associated with celiac disease; as it turned out, the fever reflected a pneumonia that had escaped notice when her pediatrician listened to her lungs. But within the past couple of years, Shira's aunt and cousins had been diagnosed with celiac disease; as long as she was going to have her blood drawn, I figured it would be worthwhile to go ahead and make sure that gluten wasn't wreaking havoc on her intestines. She is slimmer than a string bean in an extended family in which being overweight is more the norm, at least on my side. Not everyone with the condition is thin, of course, but people with untreated disease don't reap all the benefits of the food they eat due to their damaged intestines.

The joke, of course, was on me. If you look hard enough for something, you just might find it. That's exactly what happened when the phone rang on a Friday night just as we were pulling freshly baked challah from the oven. It was Shira's pediatrician reporting that her blood test had come back sky-high for the antibodies that indicate celiac disease. Celiacs must avoid gluten, which is found in wheat, barley, and rye. We had just used eight cups of whole wheat flour to bake our golden braided loaves. That night, I shared the test results only with my husband. At dinner, it was difficult to cloak my sadness as Shira drizzled honey on her challah and ate it with gusto.

A biopsy of the small intestine is the gold standard for confirming the diagnosis, and she had it done under full sedation the Monday before Thanksgiving. Her doctor called Wednesday night before the big feast, informing us that the pathologist had issued an ominous-sounding pronouncement of "complete atrophy" of the villi, the tiny hairs whose job is to absorb nutrients from food. It appeared that Shira had a case of "silent" celiac disease; blood work and biopsy confirmed the presence of disease despite her lack of symptoms. Fortunately, there's a pretty easy, if life-altering, fix: banish gluten. Abstaining from the protein causes the flattened villi to perk up again and do their thing, fluttering around in the small intestine and grabbing nutrients.

Shira's diagnosis prompted a standard recommendation, that her brother and sister and her father and I also seek testing. None of us has the disease, though we still might have changes in the family of genes known as the human leukocyte antigen (HLA) complex. The HLA complex plays a key role in helping our immune system tell the difference between our body's own proteins and those generated by insidious interlopers such as viruses or bacteria. People with celiac disease have immune systems that are inappropriately sensitized to gliadin, a gluten protein. Exposure to gluten triggers an immune attack response that causes inflammation and intestinal damage.

What's interesting about celiac disease is that nearly everyone with celiac disease possesses at least one change in their HLA genes. Yet 30 percent of the population has one of these genetic errors, and 30 percent of the population does not have to swear off gluten. In fact, only 3 percent of people with these gene changes go on to develop celiac disease, so there's clearly something else at play—an interaction between genes and environment, be it the food we eat, the air we breathe, or changes in other genes.

Celiac disease is not the end of the world. And yet, for a nine-year-old, being compelled to renounce the cupcakes brought into school to celebrate a classmate's birthday or to bow out of a pizza party is not easy. In Shira's case, finding out that she has celiac disease has been both a blessing and a curse. I know it makes no sense, but sometimes I wonder if we would have been better off not knowing now and waiting longer until symptoms may have emerged. Medically, that's irrational, but it's hard to make a persuasive case to a grade-schooler who is symptom-free about being fanatical when it comes to reading food labels, going hungry when the snack at the school play rehearsal consists of pretzels, and turning her back forever after on malted milk balls. I felt a little like Maya, Daniel's mom, saddled with worrisome medical information about a child who had seemed completely healthy.

In Shira's situation, of course, the lack of symptoms was deceiving. Though she felt fine, the protein was creating intestinal chaos. Knowing that she has an autoimmune disorder allows us to treat it. But Daniel had no special diet that could restore his genetic deletion.

⌐

People often are willing to bare their souls for meaningful change. Maya hoped that by writing a letter to CHOP, where Daniel was treated, she might influence how doctors think about genetic information and the enormous emotional punch it can deliver. At CHOP, Maya's indignant letter didn't net her a VIP (Very Insufferable Patient) award; in fact, her outspokenness paid off. For one thing, it made Sarah Noon realize that even genetic tests that are less comprehensive than sequencing need to be accompanied by detailed consent forms educating patients and their parents about the tests and what they might learn. It also scored Maya an invitation to spend an afternoon at the hospital's inaugural Clinical Genetics Think Tank—an opportunity to ruminate on how one of the country's preeminent children's hospitals had taken a wrong turn when it came to communicating testing results. At the event, held in conjunction with the University of Toronto, experts including Krantz and Spinner hashed over how to share test results, how to explain them, and, once they're delivered, how to support the families.

Maya was touched that Krantz and his team were acknowledging they needed to do better, and she decided to go. She wasn't the only parent there: Meredith Hardy, the mother of two boys also treated at CHOP, had been invited too, but for very different reasons. Hardy wanted as much information as she could get about Niels's and Lachlan's health. When Lachlan, Hardy's younger son, was three weeks old, she had to revive him after she noticed he had turned blue in his baby swing. It became routine, the resuscitation, so much so that Hardy no longer panicked when called upon to be her son's personal EMT. "I would bring my child back to life while talking on the phone with my friends," says Hardy, who is forty-five. "My kids can be fine one minute, then almost dead twenty minutes later."

Niels, twelve, and Lachlan, ten, have mitochondrial disorders, meaning their bodies don't produce energy as they should. At least, that's their working diagnosis. Mitochondrial disorders are a vast class of diseases; Hardy's sons are presumed to have some sort of "mito," as it's called, based on a muscle biopsy. Its symptoms include dysautonomia, in which involuntary functions like breathing and heartbeat and temperature control are out of whack. Medications are slowing the boys' decline but aren't stopping it. Their epilepsy, one of their most debilitating symptoms, says Hardy, "turns their brain to spaghetti."

Daniel, meanwhile, doesn't share Niels and Lachlan's struggle for survival. He's busy assembling train tracks and stacking blocks, doing the heavy lifting of a typically developing toddler, albeit one with some unexplained hearing loss.

It's a critical distinction: if you perceive your child to be healthy, as Maya did, you may not want ambiguous information that hints that he may not remain that way. If you know your child is very ill, as Niels and Lachlan are, you may have a different outlook. The theoretical balloon has already been popped; as a parent, it's in your interest to scrounge for every detail available. Hardy wants to be as prepared as possible. In fact, a study published in 2015 in *Public Health Genomics* codified the mothers' divergent degrees of interest in learning genomic information. The research, from the University of Michigan, found that parents whose youngest children had more than two existing health conditions took more interest in "predictive genetic testing"—testing to foretell future disease or disease risk—for themselves and for their youngest children.

Despite their differing perspectives, however, both mothers shared similar ideas on how results should be shared. They both wanted to caution the Think Tank participants—geneticists, genetic counselors, lab techs, and medical students, among others—to be more judicious with the results they return and to do a better job of educating patients and their parents about the scope of genetic tests.

Hardy felt starstruck upon being asked to participate. "I don't read celebrity magazines," said Hardy, whose sons are treated at CHOP by dozens of physicians. "My celebrities are doctors. For me it was like being invited to an Academy Awards pre-Oscars party." She and Maya almost didn't make the guest list. The doctors and genetic counselors had already met once—Part 1 of the Think Tank—when it dawned on someone that they were devoting hours to figuring out how to better communicate genetic complexities to parents, yet they hadn't asked for their input.

It's yet another example of how the worlds of patient and provider diverge. "These guys are getting their geek on and it's very exciting," says Hardy. "But they speak a whole different language than you and I do. If you ask me right now the difference between whole exome sequencing and whole genome sequencing, I think I forgot again. For most patient families, you need to slow it down and walk them through a counseling process."

Hardy's experience with multiple rounds of genetic testing is instructive: as she notes, genetic testing, particularly for complex, undiagnosed illness, isn't a one-shot deal. It often needs to be repeated over time to take into account new associations between genes and disease. "Genetic testing is like fishing," she says, "and what you pull out of the sea is like reading a dictionary of a newly discovered language. Sometimes you open the dictionary and the parts of speech are listed; sometimes you can read a word but not its definition; and sometimes you can read the word, the definition, and how that word is used in the language. If you think about our strand of DNA, it's like the biggest frickin' dictionary you've ever seen."

That foreign language that Hardy is referring to is one that's wholly unfamiliar to most people aside from geneticists and genetic counselors. Even other physicians who don't have genetics backgrounds are frequently in the dark. This abyss of knowledge grows wider each day with every new gene discovery. As much as sequencing is trumpeted as a miracle technology, it yields a diagnosis only about 30 percent of the time.

The process of going through genetic testing involves taking samples from blood or saliva and sometimes urine or muscle tissue. Hardy calls it "brutal—emotionally, intellectually, and even physically." Rigorous questioning and form-signing can last for hours and involve multiple family members and even extended-family members.

And it's expensive, with insurance providers covering some tests but not others. Some families have told Hardy that they thought that sequencing would be paid for, only to subsequently be socked with five-digit bills. "To go through all of this and then come up with no answers is devastating, absolutely devastating," says Hardy. "This process can be mind-blowingly fruitful or downright abusive. Genetic counselors can make all the difference."

Doctors may make the actual diagnosis, but it's genetic counselors who typically obtain informed consent and tackle education about genetic tests and what they show. But people's preferences about what and how much they want to know can change over time.

When Hardy and her husband, John Strautnieks, first had their exomes and genomes sequenced in 2009 to try to identify genetic aberrations they may have passed on to their boys, Strautnieks, who is forty-five, was adamant about one thing: he did not want to find out if he'd inherited the genetic mutation for Creutzfeldt-Jakob disease, a neurodegenerative disorder unrelated to the boys' condition. His mother had died at age fifty of Creutzfeldt-Jakob, a prion disease. Prion diseases damage the nervous system in humans and animals. In humans, the onset of Creutzfeldt-Jakob, in adulthood, is typically swift and results in death within months to a few years.

In most cases, it occurs spontaneously in about one person in a million when prion proteins go awry, creating small, spongelike holes in the brain. A different, acquired form of the disease, variant Creutzfeldt-Jakob, is related to mad cow disease; it's caused when humans consume beef from cows with prion disease.

About 10 to 15 percent of prion disease, called familial Creutzfeldt-Jakob disease, can be hereditary. This version is caused by mutations in the *PRNP* gene, which governs the production of prion protein.

Scientists aren't quite sure what prion proteins normally do, though they may be involved in protecting brain cells, or neurons, and in communication between those cells. People who have the inherited version of the disease churn out oddly shaped proteins that accumulate in the brain, killing off neurons. When those neurons die, their absence creates pinprick holes in the brain, rendering it spongelike. That sponginess causes symptoms including memory loss, fever, sensitivity to touch and sound, trouble swallowing, and seizures. And, in short order, death.

Strautnieks had never wanted to know whether he'd inherited the genetic mutation for Creutzfeldt-Jakob disease from his mother. When he learned that he could divine this information from his exome-sequencing results, he said no thanks. At the time, he couldn't handle knowing, as Hardy put it, "if he had a death sentence hanging over his head."

Several years later, as the family considered going through the sequencing process once again to see if new gene discoveries applied to the boys' condition, Strautnieks found himself changing his mind. "He would like to know," says Hardy, "because he is in an emotional place where he can take on more information."

Coming to terms with the boys' condition has made both Hardy and her husband more comfortable with the idea of peering into their future, precisely because they've realized it's going to look nothing like what they had envisioned.

"We have let go of all preconceived notions of our future," says Hardy. "We need to plan. We need to set up financial arrangements, special needs trusts. I have always wanted to know everything. If we have a good idea his life will end at fifty, we will live it up now."

She paused and reconsidered. "Actually, we pretty much already live it up now because we don't know our boys' future," she says. I asked Hardy what she meant by "living it up," and her response was stunning in its intensity and frankness:

We give ourselves permission to say "yes" as often as possible to fun, laughter, and play. We also give ourselves permission to say "no" as much as possible to anything that infringes on that quality time.

If we are on vacation, and having a really good time, if we can extend it, we will . . . Often by a week.

Yes, we play hookey! Both our boys need to be year-round learners. If we don't keep up with academic stuff, we lose traction. So, we do a lot of schoolwork in the summer. Given that, during the official school year we pace ourselves because our energy reserves are limited. We take occasional school days off to snuggle or play. On more than one occasion I have brought Niels to school late simply because Niels and Lachlan are playing together in a magical way.

No, we do not attend obligatory parental school functions and parties.

No, we do not go to church. We prefer to snuggle and watch *Sunday Morning*, the CBS show with the politics, news, and features. The boys *love* it!

No, we do not volunteer. We prefer to perform random acts of kindness.

No, we have not fixed our front door (that contains a crack through which you can see daylight). Because we decided to swim with manatees and kayak in the Everglades (near John's dad) over winter break instead.

Yes, we have spent all our savings. No, we will not be retiring.

These are the existential musings of a mother who is desperate to learn all she can about every aspect of her children's health. For Hardy, there is no such thing as too much information. Yet Maya feels equally strongly that she was saddled with results she wasn't prepared for. What if it were possible to neatly sidestep many of these paternalistic questions of access—of who owns your data, what information labs and doctors are obliged to return, and what information they're better off keeping under wraps—by flipping the paradigm? Instead of letting the decision-making power rest with the medical establishment, what if the patient's rights were paramount?

Those are the kinds of questions that Mike Bamshad is asking.

6

The Right
to an Open Future

Navigating the Return of Results

I n 2010, Mike Bamshad, the chief of pediatric genetic medicine at the University of Washington, co-led a team that was the first to sequence the DNA of an entire family—a mother, a father, and their two grown children—to find the cause of their condition. The researchers succeeded in pinpointing the gene that was responsible for Miller syndrome, a condition that caused the limb deformities and recessed chins of Debbie Jorde's two children. In accordance with standard research protocol, they engaged in the time-consuming task of obtaining formal consent from the family to share with them each potential finding. But once the results came back, the researchers were frustrated by the cumbersome process of returning those results, one by one. Complicating matters was a secondary, or incidental, finding—primary ciliary dyskinesia, a condition that resembles cystic fibrosis. As Bamshad recalls, some family members wanted some results and others wanted different ones, and some of their preferences evolved over time. Bamshad figured there had to be a more efficient way for people being sequenced to select their preferences and receive their results.

In the past, Bamshad would have performed a particular genetic test on a child and the family would come back to the clinic to get the result—testing one gene would yield one result. But sequencing scans all 19,000 or so of a person's genes. With sequencing moving toward the mainstream, the deluge of data generated from each sequencing test would mean that no provider would have the time to go over each and every potential result with each family member, asking: "Do you want this type of information? What about that type of information?"

"We can't do this," he declared at the time to his colleagues. "This is highly impractical." Together they started brainstorming solutions.

Bamshad sees exome and genome sequencing not so much as a test but more as a resource. Unlike Ian Krantz and Nancy Spinner at CHOP, Bamshad isn't conflicted about what results to share because he believes the solution is simple: make everything available, but empower the patient to decide which results to access.

That's just what patients can do through My46, a Web-based program developed by Bamshad and his colleague Holly Tabor, a geneticist and bioethicist. My46 takes its name from the number of chromosomes common to humans. Akin to an online file folder, it allows patients to manage their data, store their genetic testing results, and access particular findings when they deem it appropriate. With My46, results indicating risks for cancer or other diseases that could develop years later—like the CHOP baby's early-onset dementia gene—can be stored online in a confidential, password-protected account. Users can choose what they want to know and when. "If you are at the pediatrician's office and you are with your eight-year-old child, you can decide to pull down that result," says Bamshad. "Eventually, everyone will have their genome stored in the cloud."

For a baby being sequenced, parents can choose to learn only about results that affect their child in the near term—or opt to access everything. Meanwhile, the results remain in an online vault, where parents—or the children themselves, once they're older—can choose to tap into them at will. The tool has been used locally at the University of Washington for research results and will be licensed for free to researchers and nonprofit groups; empirical data show that families like the format. "If you generate genome information on a newborn, people wonder what to do with it," says Bamshad. "It makes no sense. We don't think it's a problem because we have this tool to deliver results."

As we observed in the previous chapter, parents' desires for genetic information about their children can vary widely. If that conversation centered around what parents want to know, this discussion explores what genetics experts and institutions think they *should* know—and how we might empower parents to take control of their children's genetic data.

Bamshad readily admits he's got a radically different mind-set from many of his colleagues around the country about divulging health data. He's not asking *if* we should be returning results; he's asking *how.* His belief that having access to genomic information early in life is useful comes from the research that he and his colleagues have conducted; from the vantage point of having se-quenced the first nuclear family, Bamshad and his team have been thinking about sequencing and the information it generates for longer than nearly anyone else. "We already have demonstrated the utility of having access to genomic information early in life, even information about a fetal genome," says Bamshad. "We've already made the decision that this is information of potential interest to families."

And yet the conventional wisdom—not to mention profes-sional guidelines—advises against genetic tests for children younger than eighteen unless they're medically necessary. "We are recom-mending that parents be left with the ambiguity that their infant is at higher risk for breast cancer because they have breast cancer in their family, but we are unwilling to test the baby to know for sure," says Bamshad, citing the familiar example of the *BRCA* genes. He disagrees. "If we're counseling someone in clinic and they are at high risk for breast cancer and they have their infant daughter with them, why should we wait to talk again when she's eighteen? Why would we not test and say she's not a *BRCA* carrier or she is, and here's what you should do about it? We should use this information to improve health care, and one of the clearest ways to improve health care is to prevent disease. It's very clear-cut."

If parents find out their child is not a carrier, they can heave a big sigh of relief. "That's huge," says Bamshad.

And, continues Tabor, who is now at Stanford University, "If she is a carrier, is it any worse to know? You're worried about it anyway. Naysayers emphasize the right to an open future: Will you think differently about your child if you know she has a *BRCA* mutation? But a mother may already treat her daughter differently because she knows that she, the mom, has a mutation in the gene. We already make assumptions about our children. There is no evidence I'm aware of that getting genetic information about your child results in a negative outcome. In most cases, there are more likely to be benefits rather than harms."

My46's role as a virtual file drawer is particularly useful in complex cases in which results are of undetermined significance. As genetic tests become more powerful, results are not always crystal clear. With gene discoveries announced regularly, it becomes even more important to store results so that they can be reinterpreted in light of new genetic insights or treatments.

It's Debbie Jorde's adult children who inspired the development of My46. Her daughter, Heather, was born in 1977, and her son, Logan, followed three years later. But it took more than three decades to figure out what gene was responsible for their condition.

When Heather was born, her hands bent down 90 degrees at the wrist; her forearms were shortened. Doctors were baffled. By the time Heather was one and a half, a geneticist told Jorde, who lives in Salt Lake City, that the likelihood that whatever Heather had would recur was less than one in a million. Still, to ease her anxiety, Jorde had an ultrasound when she was seven months pregnant with Logan. His wrists appeared straight in the fuzzy image of him in her womb, long before 3-D imaging offered realistic glimpses of that watery world. But when Logan was born, he looked just like Heather had at birth. Jorde will never forget the doctor's reaction. "The pediatrician came in and said, 'Congratulations, you just made medical history.'"

As the children grew, physicians became increasingly convinced that Heather and Logan's constellation of symptoms was genetic: their bent fingers and wrists, their cleft palates, their small forearms, their receded chins. But no one could be certain until they got the DNA proof. "For thirty-two years, we didn't know what caused this," says Jorde. Getting confirmation, in the form of a *Nature Genetics* paper unceremoniously titled "Exome Sequencing Identifies the Cause of a Mendelian Disorder," published first online in 2009, "filled a void."

After so many years, Jorde isn't unnerved by the secrets that genetics can reveal. When she, her ex-husband, and their two children were sequenced, Jorde was first surveyed by Tabor about how much she wanted to know. "We told everyone we talked to that we [were] not afraid to find anything out," says Jorde. "We've had a different life than ordinary people." What if you have a gene for breast cancer? the researchers asked Jorde. Do you want to know about it? "I said, 'Of course,'" recalls Jorde. "Not knowing doesn't mean I don't have it."

We stand at a crossroads where we must decide how to share, comprehend, and make use of information about our genomes. Forget fretting about Google knowing your shopping habits or the wisdom of accessing your bank account via the Wi-Fi connection at Starbucks: those conventional concerns about privacy pale next to the biochemical secrets housed in your DNA.

"This area is like a Rorschach test," says Robert Green, a medical geneticist at Brigham and Women's Hospital and Harvard

Medical School. "One group of people says, of course, all this information is potentially valuable to people and we should share it all. The other group is saying, don't return any incidental findings; it's inappropriate."

Trying to achieve consensus may have been a naïve undertaking, but that was Green's charge from the American College of Medical Genetics and Genomics. In 2013, an ACMG committee that he co-chaired announced with much fanfare its recommended list of incidental findings that labs across the country should always return to the patient, regardless of the reason for testing or the preferences of the family. The list was originally known as the "ACMG 56" for the number of genes for which the group said labs should scour the genome. In practice, it means that a child being sequenced, say, to help diagnose a mysterious disorder, should also have fifty-six genes scanned for changes that are associated with twenty-four or so other conditions that have potential treatments or interventions. In other words, if you wanted exome or genome testing, you had to accept the return of incidental findings from these genes.

Incidental or secondary findings, as illustrated by Daniel's genetic testing experience at CHOP, aren't new. We've all heard of cancer diagnoses made by chance when a person comes in for an unrelated scan. "If you fall off your bike and get an X-ray looking for a fractured rib, the radiologist scans the entire X-ray and automatically reports back to your doctor if something else is going on," says Green. What's changed is the scope. While a scan of an injured bicyclist could, in theory, pick up the occasional tumor, a scan of a genome, with its thousands of genes, is far more likely to pick up something suspicious.

Green's niche is examining what it means to apply genetics and genomics in the practice of medicine and more broadly in society. "Everyone will have incidental findings, depending upon what you determine to be an incidental finding," says Green, who notes that the situation is even more complicated in children. "Results that you return may never trouble the child as a child, but they could trouble that child as an adult. That then starts to get entangled with an ethical principle that's been out there for a long time that says you don't share with families medical findings if you find something that indicates a child will get cancer in their forties or fifties."

Many in the genetics community, bioethicists in particular, pounced once they reviewed the new guidelines. The stipulation that patients must be tested for certain conditions, whatever their wishes, especially troubled them. "It's not 'incidental' if you have a mandatory hunt," says Lainie Friedman Ross, a pediatrician and bioethicist at the University of Chicago.

It's little surprise that Ross disagreed so vehemently with the ACMG recommendations. Around the same time the ACMG 56 guidelines were announced, Ross had authored an eagerly awaited policy issued jointly by the ACMG and the American Academy of Pediatrics, the nation's professional association for pediatricians, about genetic testing in children. The statement differs rather starkly from the ACMG 56 guidelines, even though the ACMG was one of two parties to its development. Casting a wide net for potential disease-causing gene changes is not advised; for the most part, genetic testing should be performed only for diagnostic reasons, because testing should be in a child's "best interest."

What's tricky is that not everyone agrees on how to define that concept, especially since what's in the best interest of the child may, in a broader context, encompass what's in the best interest of other relatives. The ACMG recommendations, recall, direct labs to scan children's genomes for a multitude of disease-linked variants that parents did not request. As a result, not only could disease or risk for disease be caught early in a child, but it could be identified in an unsuspecting adult as well. That theoretical *BRCA* mutation that Bamshad and Tabor referenced may be of little consequence for a young girl who is years away from developing breasts. But it could be of grave consequence to her mother or father, who is likely to have passed it on to her. "Especially if it is the mother, this information could be life-saving," says Green. "There is no way that losing a mother to breast cancer is in the best interests of the child."

Be that as it may, requiring that kids be automatically tested for disease risk feels coercive to others. "A typical pediatric ethics framework [asks what is in] the best interests of the child, but here it's about helping the family members," Ellen Wright Clayton, the Vanderbilt University pediatrician and bioethicist, told a small group gathered for a 2013 bioethics seminar at Seattle Children's Research Institute. "In other settings, I have talked about the child as a canary in a coal mine here. It feels that way to me. From a pediatric ethics perspective, [the ACMG recommendation] is a stunning deviation from what pediatric ethics has been based on."

Whether you agree or not with the marching orders to look for additional findings, Ross raises a good point: Can something rightfully be considered "incidental" if it's intentionally sought out?

In 2014, the ACMG officially concurred. The organization now refers to "incidental findings" as "secondary findings," which is the term I'll use when appropriate throughout the remainder of this book. The geneticist Sherri Bale, who served on the ACMG committee that made the decision, believes the distinction is important and more than just a matter of semantics. "We are looking for this stuff," she says. "It's not incidental. So let's just call a spade a spade."

With that hurdle out of the way, debate has continued to percolate within the genetics community over which genes, if any, should routinely be scrutinized. (The list of genes recently expanded to fifty-nine.) After considerable uproar, the ACMG revised its guidelines in 2014, backpedaling to say that parents—indeed, anyone—may opt out of additional genome screening. If you don't want surplus screening, you won't automatically get it.

❑

Spend time at any meeting of a genetics or bioethics professional association, and you'll notice lots of chatter about ELSI. The first time I heard the name, I assumed it was a female researcher at the conference, some eminent scientist brandishing an impressive stack of published papers. In fact, ELSI is an acronym for the ethical, legal, and social implications of the emerging world of genomics that we find ourselves navigating.

In October 2014, the American Society of Human Genetics and the American Society for Bioethics and Humanities held a joint symposium that focused on the enigmatic ELSI. Robert Klitzman, a psychiatrist who runs the Master of Science in Bioethics program at Columbia University, proposed at the meeting one way of thinking about what information should be shared: "preventing serious harm to another person may at times outweigh a person's right to confidentiality." In other words, if genetic information stands to significantly impact another person's health, it's not necessarily okay for providers to keep quiet. As support, Klitzman—who has written a book, *Am I My Genes? Confronting Fate and Family Secrets in the Age of Genetic Testing*, about how people decide whether to pursue genetic testing—resurrected a storied legal case in the annals of mental illness: *Tarasoff v. Regents of the University of California.*

It was the fall of 1968 when the graduate student Prosenjit Poddar met Tatiana Tarasoff at a folk dancing class at the University of California, Berkeley. On New Year's Eve, they shared a kiss. Poddar interpreted it to mean that they were an item; Tarasoff didn't share his conviction. Under normal circumstances, Poddar would nurse his feelings of rejection, shed a few lovelorn tears, and then move on. Instead, he became a stalker. He was inconsolable and depressed for months and began seeing a psychologist, in whom he confided his intention to kill an unnamed woman, presumed to be Tarasoff. The therapist told campus police that Poddar should be committed to a mental hospital; he was briefly taken into custody but released after law enforcement concluded he was rational. His psychologist's supervisor said there was no need for Poddar to be further detained. But Poddar clung to his obsession with Tarasoff; he even buddied up to her brother and moved in with him while she was visiting family in Brazil over the summer. Then, in October 1969, more than nine months after their kiss, Poddar stabbed Tarasoff to death. Despite Poddar's admission to his psychologist, Tarasoff was never warned of his intentions.

Her death prompted her parents to sue the Regents of the University of California. The trial court dismissed the case, but on appeal, the Supreme Court of California sided with the family. The upshot: in that state and those that follow its ruling, therapists have an obligation to warn or protect intended victims of serious crimes so that they may take appropriate precautions.

What does this have to do with genetics? Well, perhaps, says Klitzman, it's not such a huge leap from a therapist's duty to prevent harm to a physician's duty to do the same by disclosing potentially serious and treatable findings. But what qualifies as serious enough, and who should decide? In fact, the law has wrestled with this issue. In *Safer v. Estate of Pack*, a New Jersey court ruled that a physician (Dr. Pack) had a duty to take "reasonable steps" to warn a patient's relatives that they may be at risk for colon cancer.

Consider how things could have turned out had Laurie Hunter, a high school English teacher from Jackson, New Jersey, not learned as a result of her daughter's genetic testing that she too has an increased risk of cancer.

A mother of three, Hunter has two daughters with severe but unrelated genetic conditions. Hunter discovered that she is at risk for developing cancer after she was found to have the same genetic deletion as her teenage daughter, who was being tested to determine why she was developmentally delayed, her arms so stiff that she couldn't wipe herself, her muscles so weak that she couldn't blow her nose. Hunter was stunned by the news: the average age of onset for the tumors to which she's predisposed is thirty; she was forty-two when she learned she is missing the same seven genes on her first chromosome as her daughter Amanda. She is now undergoing regular scans to detect any potential cancer as early as possible.

Hunter first suspected something unusual was going on when Amanda, now seventeen, was just two months old. She was floppy, like a sack of potatoes—what clinicians call "low muscle tone." Amanda, with straight, dark hair and dark eyebrows arched above dark eyes, didn't walk until she was two years old. She's been in speech and occupational therapy her entire life, and, says her mom, has an IQ of 75 to 80 "on a good testing day."

As it turned out, Amanda's missing genes don't appear to explain her delays, but one of them, *SDHB*, is associated with cancer. *SDHB*—succinate dehydrogenase complex, subunit B—is a tumor suppressor gene; its absence means it can't do its job suppressing tumors. A change in the gene—or in Amanda's case, a deletion—results in an increased risk of a type of cancer called hereditary paraganglioma–pheochromocytoma syndrome type 4. Tumors typically pop up in the abdomen, but they can also develop in the head and neck.

To determine if the deletion had occurred spontaneously or had been inherited, Hunter and Amanda's father, who are divorced, were both tested. When the results came back that Hunter shared the same deletion, a genetic counselor phoned with the news. Hunter was shocked. "It's like your worst nightmare coming true, to find out you carry something as a mother and you have passed it on to your child," Hunter wrote in a first-person essay that accompanied a series I wrote about sequencing children's genomes.

Considering the context of Hunter's life, the blow is even harsher. Hunter's other daughter, Kailyn, born in 2010, was diagnosed with a random, or de novo, deletion on her fourth chromosome; she has Wolf-Hirschhorn syndrome and is even more disabled than Amanda. Hunter's middle child, Ryan, who is two years older than Kailyn, is developing typically. And therein, almost counterintuitively, lay the worst part of learning that Amanda's deletion had been inherited from Hunter. If Hunter had passed it to Amanda, she could also have passed it to her other children. "As soon as the genetic counselor told me that I carried the deletion, I immediately knew my son [was] also at risk. That is the part that devastated me. I know this sounds bad, but my daughters already have so many things they suffer from that going to an oncologist is just one more doctor. Now we have to send my son into that mix. It is a new level of anxiety. I have this one child where all my dreams lie and now he may have this too."

Hunter breathed a huge sigh of relief when she learned that Ryan doesn't carry the deletion that she and Amanda have. But life remains incredibly complicated.

> I still have to worry about myself now. Because of these two kids with extraordinary medical conditions, I don't have a life. I can't remember the last time I had a regular physical. I could have blood pressure through the roof, and I don't know it. On Mondays and Wednesdays, I go straight from work and pick up the girls for 2½ hours of occupational, speech and physical therapy. Tuesday is horseback therapy. Thursday is my open day so if I need to take them to doctors, that's when I do it. Friday is for my son. I signed him up for gymnastics because he's been shuffled around to doctors and therapy for so long.

Now, in addition to blood work for Amanda once a year and full-body MRI scans every other year, Hunter needs MRIs, too. One discovered a lesion above her diaphragm. She wrote in a wry email:

> They think it is a paraganglioma tumor. I went last Thursday for a PET scan, and I am hoping to have results tomorrow. If it is, indeed, a tumor, I will need surgery over my winter break . . . sigh.

What if a decision had been made not to reveal Amanda's genetic deletion and its potential implications to Hunter on the grounds that the average age of onset for cancer was thirty and Amanda was just a teenager at the time—not to mention that the deletion did not provide an explanation for Amanda's delays? It's possible that Hunter might have developed a tumor that was caught only in an advanced stage because she would not have known to request preventive screening.

How preventable or treatable a condition is determines what the medical community calls "actionability." Can you do something about a diagnosis? Can you take action? If the answer is yes, doctors support testing children. If the answer is no, they usually don't.

Long before sequencing became an invaluable component of health care, Marcia Van Riper, a nursing professor at the University of North Carolina at Chapel Hill, decided to craft a research study that would examine families' experience of genetic testing for five conditions, including Huntington's disease, which occurs as nerve cells deteriorate in the brain. Huntington's is frequently held out in genetics as the classic dread scenario: if you have the genetic aberration—a series of three DNA bases that repeats dozens of times, like a vintage record skipping—you are bound to get the disease. In a normally functioning gene, the sequence—CAG—mirrors itself between eleven and twenty-nine times, but a faulty gene repeats the sequence up to eighty times. That prompts the protein huntingtin, which is produced by the *huntingtin* gene, to misfold and tangle, making it cluster in the brain and kill nerve cells, especially in the basal ganglia, an area of the brain that directs movement, and in the cortex, which oversees thought and memory.

Huntington's occupies a unique place in genetic history. In 1983, the gene became the first to be isolated to a specific spot on a specific chromosome; ten years later, the actual gene mutation was pinpointed.

Symptoms of Huntington's typically start appearing in middle age, often after prime childbearing years. Medication can ease some symptoms, but no therapy can alter the fatal course of the disease.

If you have the disease, each of your children has a fifty-fifty chance of developing it as well. Unless something breaks the cycle—say, all affected people in a family choose to adopt, or decide not to have children; or, alternatively, opt for preimplantation genetic diagnosis, allowing would-be parents to select an embryo that does not have the mutation, or prenatal diagnosis, followed by abortion if the fetus does have it—the disease will continue to clamber up a family tree, weakening roots and snapping branches.

Amid the devastating narratives of families coping with a diagnosis, Van Riper came across a striking story of renewal: surviving spouses in two families struck by the disorder fell in love. In one family, the husband had died of Huntington's; in the other, the wife had succumbed. "Someone in the clinic thought it would be nice if they met," says Van Riper. "They had ten children between them, and the surviving spouses met and got married."

The children ranged in age from ten to twenty-one when their parents got married. By the time Van Riper interviewed them, some of the children—all of whom were grown by this time— had already been diagnosed with the disease or knew they had the gene and would develop symptoms; it was only a matter of time until they started becoming uncharacteristically irritable or depressed or struggled to remember a friend's name or respond to a question.

Typically, the disease unspools over fifteen to twenty years. People affected might experience uncontrolled movements called chorea, clumsiness, and problems with coordination. They might stumble and appear drunk. (Woody Guthrie, who inherited the disease from his mother, was misdiagnosed for years, his unpredictable jerking and botched lyrics attributed to too much alcohol. Once he was finally diagnosed in the early 1950s, he wrote a poem, "No Help Known," whose meaning remains as poignant today as it did then: "Huntington's Chorea / Means there's no help known / In the science of medicine / For me.")

It's an agonizing death as skills like walking and talking decline and swallowing becomes difficult. In its later stages, Huntington's can resemble Alzheimer's, as some people no longer recognize relatives.

In the blended family that Van Riper interviewed, three of the ten children had already died, three were showing symptoms of Huntington's, two had tested negative, and two had chosen not to be tested. "What was interesting," recalls Van Riper, "was the ones who had been tested felt the ones who hadn't been tested were being selfish because their children had a right to know."

Before the gene was mapped, when the ability to learn your fate was still theoretical, about three-quarters of at-risk people said they would want to know their status. Once the gene was mapped, it spurred the development of a DNA test to assess the number of CAG repeats: an expanded repeat of more than forty CAGs in a row indicated that the disease was waiting to strike. The test was simple—just a blood draw—but now that it was available, researchers discovered that the majority of people who had said they wanted to know if they were destined to get Huntington's had dwindled to about 25 percent.

Before a direct gene test was available, however, a "linkage" test was rolled out. This initial testing for Huntington's began in 1986 as part of a research study at Johns Hopkins University and at Massachusetts General Hospital in Boston. Because the actual gene had not yet been firmly identified, the test consisted of "markers" that were linked to the gene. The protocol for testing presymptomatic at-risk people was carefully constructed: six visits, including four counseling sessions; after the fifth visit, testing would take place. At the final visit, the results would be delivered. Participants were required to have a partner accompany them. The protocol that this study used became the gold standard for presymptomatic testing for Huntington's.

When Jason Brandt, a Hopkins psychologist, had asked the patients being tested about their feelings regarding childbearing, one prevailing theme was that having kids was central to living a worthwhile life. Today, attitudes toward having children is one of the themes that Debra Mathews is examining. Trained as a geneticist, she focuses now on ethics and policy considerations and moves fluidly between her fields at the Berman Institute of Bioethics at Johns Hopkins. "With Huntington's, it's a worst-case scenario: we tell someone how they will die," she says. "It's relatively rare but incredibly devastating. It's such a dramatic disease that our early experience with it—because it was the first gene to be mapped—has shaped our thinking around the return of genetic results."

Mathews is revisiting the same group of people tested in the original study, the partners who accompanied them, and their adult children, who would have been very young or not even born when their parents were tested. Did the research participants end up having children? If so, what do those adult children think about the decisions their parents made to get tested and to have kids? And how are adult children making their own reproductive decisions based on technologies available today? Mathews and her colleagues are particularly interested in those adult children whose parents tested positive—meaning their testing revealed an expanded repeat, as opposed to a normal repeat—and went on to conceive and pass the genetic signature on to their kids, like an exceptionally unwelcome family heirloom. In short, how do families cope?

"Jason's research with the early cohort suggested that they did pretty well, and that folks who were found to have an expanded repeat sometimes did better than those who had the normal repeat in terms of depression and adaptation," says Mathews. "The vast majority of folks do fine integrating this into their lives. There are cases of folks with Huntington's committing suicide after learning of their diagnosis, but most people simply move on with their lives. This experience has loomed so large in our thinking, and in many ways we've taken the wrong message from it, that people can't handle truth, that they can't handle genetic information about themselves. Our data is bearing out that this is not true."

Knowing that Huntington's lies in wait allows a person to make more informed personal and professional decisions. If you knew you'd get sick in middle age, maybe you wouldn't pursue a graduate degree that requires years of schooling and training. Maybe you would decide to conceive a child much earlier than you would have otherwise, or maybe you'd use reproductive technology to avoid having an affected child.

Knowing that you won't develop the disease, meanwhile, eliminates years of uncertainty, but it may also introduce guilt if, say, you are negative but a sibling is positive. For many, remaining in the dark seems to be the most palatable choice, choosing the anxiety of continued uncertainty over potential emotional upheaval.

Everyone tolerates anxiety differently, a fact acknowledged in the genetic testing guidelines from the American Academy of Pediatrics and the American College of Medical Genetics and Genomics. While they now support testing children at any age for disorders that present in childhood, they continue to oppose testing children for a disease such as Huntington's that may develop in adulthood. The new recommendations do allow for exceptions in certain circumstances. For example, consider a family that has weathered generations of breast cancer diagnoses. Previously, the medical establishment had been fairly black-or-white about testing kids for diseases that wouldn't appear for decades. The prevailing wisdom? Don't do it. But the new philosophy holds that testing can proceed in limited circumstances, such as if the family history is causing great anxiety and the parents and child—an adolescent, ideally—each support testing.

Indeed, the writer Judith Rosenbaum explains in an essay in the online Jewish affairs magazine *Tablet* that she was five years old when her mother was diagnosed with breast cancer. Despite five recurrences in thirty-three years, her mother didn't want to get tested for a *BRCA* mutation. Rosenbaum was similarly resistant until she had her own children, boy and girl twins. She tested positive for a mutation and elected to have surgery to remove her breasts and ovaries.

Her children, toddlers at the time, were "blissfully ignorant of the family legacy that lingers silently in my genes, and possibly in theirs." But by the time Rosenbaum's daughter turned seven, she asked if she too would get breast cancer. That, of course, is a question that Rosenbaum couldn't answer definitively. She wrote: "I imagine that all parents eventually must come to terms with the fact that the legacy they've passed on to their children is in some way mixed—a blend of positive and negative characteristics, emotional and physical. Our case is just clearer, documented by a genetic test and several scarred bodies."

Rosenbaum did not seek to test her daughter at such a young age. Regardless, it's doubtful that a doctor would agree to test a seven-year-old. According to the professional guidelines laid out above, teenagers have more standing than younger children. If a teen requests testing, her wishes may prevail. Similarly, if a teen has no interest in testing but her parents support it out of fear that their child may have inherited a mutation that raises the risk of developing cancer in adulthood, Mom and Dad are out of luck. But the ball bounces back to the parents' court if the disease in question manifests in childhood: "In the case of predictive testing for childhood-onset conditions or conditions for which childhood interventions will ameliorate future harm . . . parental authority to determine medical treatment supersedes the minor's preferences with regard to liberty and privacy."

The increased nuance and flexibility reflected in the guidelines indicate that doctors and ethicists recognize that genetic data can be helpful. Knowing that a child is on track to develop a disease can help a family to plan—allowing them to research specialists or, for that matter, relocate to a ranch-style house if they expect their child to be confined to a wheelchair.

Yet there are serious concerns about how genetic information may erode a child's right to an "open future"—one unfettered by the knowledge of potential medical complications. It's being debated in all aspects of genomic medicine now, but particularly as it relates to sequencing babies. Jonathan Berg, a geneticist at the University of North Carolina at Chapel Hill who is researching the implications of sequencing newborns, thinks that sequencing an infant's genome, then dumping all the data in the parents' lap, is problematic because all of the information isn't equal. To illustrate his point, he grabs a piece of lined notebook paper from his desk in a sunlit office in the tawny brick Genetic Medicine Building on the UNC campus and speedily sketches out a graph as I watch, then divides it into quadrants and assigns each quadrant a category of disease. It's clear he's given this more than a little thought.

Here's what it looks like:

- Upper left: "childhood-onset" conditions (those that appear before a child turns eighteen) that can benefit from treatment
- Lower left: childhood-onset conditions that have no treatment or cure
- Upper right: adult-onset conditions that can be treated or mitigated with preventive measures such as a prophylactic mastectomy
- Lower right: adult-onset conditions that have no treatment or cure

If Berg could design an ideal genomic screening paradigm for healthy people, it would go like this: everyone would be given upper-left results for childhood-onset conditions that can benefit from treatment; parents could request lower-left results for childhood conditions that have no treatment; anyone could receive upper-right or lower-right results about adult-onset conditions once they turn eighteen.

Every day, of course, parents influence their children's health. They decide what to feed their kids. They decide whether to require them to wear helmets when riding their bicycles. They choose whether or not to smoke in front of their children. "But would you have wanted your parents to decide for you about which genetic information to learn?" says Berg. "Or would you say, 'That's none of your business; it's my genome?'

"If they decide to learn about your genome, if parents have already opted to learn about adult-onset conditions and it's in your medical record, it's a decision that's been taken away," Berg told me as we left his office together, squinting in the late-afternoon sunlight. He was on his way to pick up his daughter at preschool. "The right to an open future has been taken away from you, and that's a very powerful argument in my mind as to why genetic testing should be limited in children."

❒

The concept of the right to an open future can be traced to Joel Feinberg, a legal philosopher at the University of Arizona who in 1980 argued that children have special "rights-in-trust" that they simply can't take advantage of as minors. These rights differ from the rights of adults, he explains, because "the child cannot very well exercise his free choice until later when he is more fully formed and capable." To illustrate what qualifies as a right-in-trust, Feinberg employs an extremely graphic but highly effective analogy: take, for example, "the right to walk freely down the public sidewalk as held by an infant of two months, still incapable of self-locomotion. One would violate that right in trust *now*, before it can even be exercised, by cutting off the child's legs."

Amputating those baby legs (perish the thought) is the equivalent of foreclosing on that child's right to an open future, curtailing his rights before he is able to exercise them. "His right while he is still a child is to have these future options kept open until he is a fully formed, self-determining adult capable of deciding among them," writes Feinberg.

In other words, if parents decide to perform a bunch of genetic testing on a child not old enough to make a decision for himself, they have taken away their child's right to decide whether he wants that testing in the future. Parents are infringing upon the right of their child to later self-determination.

There's been no shortage of chagrin within the medical community about the potential for psychological harm in the face of disconcerting test results. Experts worry that fear that children will get sick will prompt parents to treat those children differently than their presumably healthy siblings and hinder parent–child bonding. With that in mind, is it appropriate for doctors to withhold some of these secrets from parents? If you knew your child would develop dementia, would you treat her differently? Love her less—or more?

In medicine, there's a long-held belief that knowing about disease or disability impairs family bonding. More than fifty years ago, Drs. Morris Green, a pediatrician at Indiana University, and Albert Solnit, a pediatrician at Yale, described in the medical literature a case study of twenty-five children who were so sick or injured that their parents had expected them to die. Fortunately, the children defied the dire predictions, but the ending didn't track the happily-ever-after of a classic Disney plot. It was more like the Viennese saying that Green and Solnit used to preface their research: "So many people die who never died before." The children, they observed, displayed disturbances in their psychosocial development and in their relationships with their parents, including "difficulty with separation, infantile behavior, bodily overconcerns, and school underachievement." Green and Solnit coined a term for this behavior. They called it "vulnerable child syndrome," and the name has endured.

"Vulnerable children" have parents who still consider them at risk, maybe even on the brink of death, after they recover from illness. The clinical features of this syndrome, wrote the doctors in 1964, are such that "the parents have the feeling that these children are not completely theirs but only on a tenuous loan." A mother may experience prolonged depression "during which her ability to relate warmly and intimately with the child becomes markedly impaired." Others say they have recurring nightmares about their child dying. Still others wake up multiple times a night to check that their child is alive. (This last observation did not strike me as incredibly pathological; even though two of my children are in double digits, I still tend to plant a kiss and place a hand on their slumbering backs to feel them rise and fall before I turn in most nights, just to check that all is right.)

Parents of so-called vulnerable children in the study also reported being overly permissive with their parenting practices, although in truth, their behavior doesn't sound all too different from today's helicopter parents. "The parent is overprotective, overly indulgent, and oversolicitous while the child is overly dependent, disobedient, irritable, argumentative, and uncooperative."

Years later, when the pediatrician Jack Shonkoff wrote a commentary on the original paper's "seminal creativity and its enduring salience over more than 3 decades," he noted:

> The most important contribution of this report is the extent to which it underscores the critical importance of what physicians say (and don't say) to parents. The lesson for the clinical community is powerful and clear. Anything that a pediatrician conveys to a mother or father, whether it is thought through clearly or not, can have enormous impact on them and their children. Moreover, the effect may not only be immediate, but it often can be long-lasting. Comments made at particularly sensitive moments are particularly potent—the initial communication of the diagnosis of a serious disease or disability; a casual comment about the potential implications of a presenting sign or symptom; a spontaneous remark about the severity of an acute illness; an off-the-cuff response to a parent's question . . . Even the most conscientious physicians cannot remember everything they say to parents in the course of a busy day. Yet, an especially poignant remark

about a child, whether delivered deliberately or casually, is likely to be remembered and quoted verbatim by the parent forever.

Decades after a doctor's cutting remark, Carolyn Hintlian, for one, still recalls how much it hurt to hear her son's doctor predict that James, who has Down syndrome, could aspire at most to "make a peanut butter sandwich."

Nowadays, vulnerable child syndrome is increasingly invoked when referencing potential harm caused by sharing uncertain genetic information about children with their parents. It's one of the primary reasons that some physicians and bioethicists advocate returning only some and not all results to parents. It's also why some experts urge caution before throwing the door wide open and expanding the number of diseases tested for in newborns. The joint pediatric and genetic guidelines that Ross authored had warned that doing so could create a generation of " 'patients in waiting': individuals with a genetic diagnosis who have no signs or symptoms and may remain asymptomatic for years or decades."

Robert Green, who helped develop the ACMG 56 guidelines, believes that concern about too much information contributing to parental anxiety, distress, and misunderstanding is largely speculative. "There are all these theories about harm," says Green, who has had his genome sequenced, along with his wife's, and has had his three children's DNA analyzed by the personal genetics company 23andMe. "People learn bad news all the time and it is upsetting. But what if you learn bad news that you can actually do something about?" He referenced actress Angelina Jolie's decision to publicly share that she'd chosen to have her breasts and ovaries removed after she found out that she has a *BRCA* mutation that ups her risk of cancer.

"The right to an open future is a ridiculous argument," says Green. "I suppose it's possible a particular family or child could experience psychological distress from learning about cancer risk, but that's a theoretical harm. We know that the damage to a child from losing a mother or father is a definite harm. Say that through a child, a sudden cardiac death mutation is identified in a parent. How great would that be for the kid?"

Even Ross, author of the guidelines, has acknowledged that decades of follow-up work have revealed that concerns over vulnerable child syndrome may be overblown. "We still worry about that, but we've come to realize to the extent that there are data that the concerns were probably overhyped," Ross told me.

Joanna Fanos has tracked this phenomenon of hype and hyperbole. In a decades-long career studying the impact of serious pediatric illness on families, she's actually found that the knowledge that a child is sick or at risk of becoming sick may intensify the parent–child bond, particularly between mother and child. Treatment demands extra time and attention. Far from eroding the bond between parent and child, Fanos found that a childhood diagnosis may cause parents to ramp up the time they spend with an ill child in what is called "anticipatory mourning"—the process of preparing emotionally for what may be a shortened life span. In fact, parents may spend so much time focusing on their sick child that they neglect their other children, a consequence that Fanos has chronicled. "If parents have a few kids, they'd say, 'I'm going to spend all my time and emotional energy on Johnny because my other kids will be around forever," says Fanos, a research psychologist at Children's Hospital Oakland Research Institute.

Maya Hewitt chose not to contend with that possibility. As you may recall, Maya is the mother of Daniel, the tow-headed toddler whose less-than-perfect hearing netted him a trip to the genetics clinic, which turned up a secondary finding—a gene deletion whose uncertain prognosis has caused his parents no small amount of worry.

When Maya and Daniel's father, Andrew, had themselves tested, they waited months for the results—one of the factors contributing to Maya's angry letter to her doctor. Finally, in the summer of

2014, she, Andrew, and Daniel were in the car heading back from a vacation in Maine when Maya's cell phone rang. It was the genetic counselor with the results of the testing that had been done on Maya and Andrew.

"Andrew's results are normal," the counselor said, "but you have the same deletion that Daniel does."

Later, Maya spoke with a specialist in dyskeratosis congenita, a disorder that can be associated with her family's missing genes. The specialist told Maya that she believes that both mother and son are silent carriers and that there are no indications that Daniel will go on to develop a severe form of the disease, although there's some research that the condition can worsen with each subsequent generation. While Maya found that heartening, it mostly served to reinforce Daniel's status as an only child. "We always thought we wanted to have one child and do it well," says Maya. "But now knowing this, it certainly solidifies that decision. We're open if something changes, but we don't want to have another child who could potentially live with this."

While some parents wrestle with unwelcome genetic information, those who've long been tormented by an absence of answers are often grateful for any and all knowledge. Parents of kids with undiagnosed disease, in particular, crave any snippet of data, any potential explanation for why their child has fallen off the developmental rails. Genome and exome sequencing, tests that peer deeply into DNA, have the potential to put a hard stop to their family's diagnostic odyssey.

7

How to Hunt a Zebra

Ending the Rare-Disease Diagnostic Odyssey

I n medical school, students learn the art of diagnosis. Rule
number one: start simple by thinking first about common
diseases. If a patient presents with a cough, a doctor is trained to initially suspect an upper respiratory infection, not lung cancer, as the
most likely offender. Sometimes, of course, that approach leads down
the wrong path. But without it, our medical system would grind to
a halt. Even the most straightforward diagnosis would involve numerous costly tests and workups, the equivalent of medical multiple choice, ranging from the most plausible of possibilities to the
most unlikely. "If you hear hoofbeats, you think horses, not zebras,"
explains Shelly Bosworth, who works at a company that inverts
that paradigm. Bosworth is a genetic counselor at GeneDx, which
specializes in rare-disease diagnosis. The zebra is the company's
mascot because "we do the opposite—we think zebras, not horses."

The zebra has become so emblematic of the quest to identify
elusive diseases that GeneDx has commissioned squeezable foam
facsimiles to give away as swag at genetics conferences. I have a
red-and-white one on my desk—presumably an even rarer itera-
tion of the black-and-white savanna dweller.

For the longest time, geneticists encountering patients who presented with mysterious symptoms were left with few navigation tools to assist with their zebra tracking. After a thorough exam that yielded few clues, they'd have little choice but to play what amounted to a well-informed guessing game, testing first for one suspected condition, then, when that proved fruitless, testing for another, and so on. The process was frustrating for doctors and their patients, more often than not young children whose parents had noticed they were not meeting their developmental milestones— rolling over at a few months old, sitting at around six months of age, crawling or scooting not long after. Hit-or-miss testing was costly and inefficient, and not infrequently, it turned up no answers.

This guessing game, in many cases, no longer exists. The ability to sequence DNA has opened up a world of diagnostic possibility. The Duke geneticist Vandana Shashi calls sequencing the "first technological advance we've had in decades," as it helps researchers match symptoms with a cause and target culprit genes that have never before been associated with disease. For sick children whose diagnoses have eluded doctors, sequencing is a godsend. This comprehensive approach has drastically curtailed what for many families was an arduous sleuthing mission that could extend for years and years. Amid a cascade of concern about genetic testing and its appropriate limits, sequencing in cases of rare disease has emerged as one of the most helpful and least controversial uses of this powerful technology.

In 2012, Shashi and David Goldstein, a geneticist who was director of the Duke Center for Human Genome Variation, published a pivotal paper showing that sequencing could be successfully employed as a biochemical Sherlock Holmes when doctors were stumped by patients with mysterious diseases thought to have a genetic basis. A condition is generally considered to be genetic in origin if multiple family members are affected, if the condition appears likely to be a result of abnormal development, or if similar conditions are known to have a genetic cause. For example, severe intellectual disability that's not related to a birth injury or an environmental toxin is very likely to have a genetic component.

In truth, most conditions are caused by the interplay of genes and environmental exposure. "People tend to think of genetics as fixed, and that's a problem," says the University of Washington's Mike Bamshad. Just because a disease is genetic in nature—caused by a change in a person's DNA code—it doesn't mean that there aren't other contributing factors as well. "Genetic versus nongenetic is an easy way to categorize things, but it's not an accurate reflection of reality," says Bamshad. Even so-called single-gene disorders can be multifaceted. Take cystic fibrosis, for example. A mutation in the *CFTR* gene is necessary and sufficient to cause disease: if you have a *CFTR* mutation, you have cystic fibrosis. But the severity of your experience with the disease draws on other factors. People with cystic fibrosis, for example, are at greater risk of getting sick from *Pseudomonas aeruginosa*, a rod-shaped bacterium that causes serious lung infection. They might also have changes in genes that control the immune response to that pathogen, potentially worsening their reaction.

A rare disease doesn't have to be genetic in origin, but most probably are, researchers suspect; indeed, they believe most rare conditions are most likely single-gene disorders. As rare-disease advocates point out, the term "rare disease" is a misnomer. When considered on an individual basis, rare diseases occur infrequently. In the United States, a rare disease is defined as a condition that affects fewer than 200,000 people, a definition that was articulated by Congress in the Orphan Drug Act of 1983. But tally these rare diseases up and the numbers are overwhelming. There are 30 million people living with rare diseases in the United States—that's one of every ten Americans—and another 30 million in Europe. Globally, rare diseases are thought to affect 350 million people. To put that number into perspective, Global Genes, a rare-disease advocacy organization, points out that if everyone with a rare disease were to take up residence in a single country—and ostensibly boot out their healthy compatriots—this genetically compromised territory of Raredisease-istan would claim the position of the world's third most populous country. To make matters worse, less than 10 percent of rare diseases have an FDA-approved drug treatment.

Linking a gene with a particular disorder aids in diagnosis. But it goes beyond that. Learning more about the correlation between genes and disease also offers insight into how genes work, which in turn can spark ideas for new therapies and treatments.

"In a perfect world, we would find a mutation and there would be an FDA-approved treatment," says Misha Angrist, the Duke researcher who was among the first people to have his whole genome sequenced. "But most of the time, that's unlikely to be true, that we will be able to have some miraculous effect on their treatment. This is really about trying to let parents avoid diagnostic odysseys, bouncing around from doctor to doctor, being told different things, people shrugging their shoulders. At least, we want to be able to say, 'Your child has a problem with his blood-clotting system and that can be traced to a defective gene.'"

❐

Homer's *Odyssey* recounts a long and perilous journey beset with shipwrecks, a menacing Cyclops, a band of cannibals, and assorted acts of murder and mayhem, all of which Odysseus weathers on his journey home. It takes ten years for our hero to reach his destination. For some sick children like Adam Foye, their "diagnostic odyssey"—what doctors term the months that stretch into years as a person lives with an illness that defies definition—lasted even longer.

Adam's diagnosis took more than a decade because he was born before sequencing was available. Adam, in sixth grade, couldn't walk down the grocery store aisle unassisted; he had missed sixty days of school as a fifth-grader. When he was eleven, researchers were finally able to use sequencing data to link his extreme muscle weakness with a mutation in the *TTN* gene, which encodes a large muscle protein called titin. His mother baked a cake and frosted it white, and Adam shakily wrote "TITIN" in yellow icing. It was cause for celebration even though his diagnosis of titin-related centronuclear myopathy had no known treatment. "As my husband says, this is not our final destination on our medical journey, but it's an important milestone," Sarah Foye said in 2012, soon after learning the news.

Even though there's currently no therapy, let alone a cure, identifying Adam's errant gene is a step on the path toward treatment. Only once a particular gene change is linked with disease can researchers start working on precise treatments; up until that point, there's no gene on which to focus their efforts. "Until you find the part in the entire machine that is altered or broken, you don't have any reasonable hope of trying to fix it," says David Margulies, executive director of the Gene Partnership at Boston Children's Hospital, which helped sponsor a gene-sequencing competition that diagnosed Adam. "Once you've identified the part that is altered, you don't have any assurance that you can fix it, but at least you have a starting point."

Like the Foyes, families who've long been seeking answers are astounded that this new technology is now often able to yield a diagnosis where years of previous analyses that tested genes or a portion of genes individually could not. Adam's narrative of going years without a diagnosis until sequencing solved his medical mystery is now fairly common.

What's not as typical—yet equally compelling—is the story of tiny Cara Greene. When doctors speak of using sequencing to end the diagnostic odyssey, there's no better evidence of its potential than her experience.

Wispy-haired, pixie-faced Cara Greene was fifteen months old in November 2013 when she got sick with a fever that lasted for five days. Shortly after it resolved, Kristen, her mother, prepared to take Cara to the pediatrician for her fifteen-month well child check-up. Cara's father, Clayton, is trained as a physical therapist, and it's that background, that awareness of the physical self, that may have alerted him that Cara's eyes seemed off. It was hard to describe, but it was almost like they were bouncing up and down—a condition that is called nystagmus. Ask the doctor about her eyes, Clayton told Kristen, a lawyer who stopped practicing shortly before Cara got sick. The pediatrician observed Cara, then referred her to a neurologist. It's probably nothing, Kristen remembers the doctor saying; possibly a brain tumor, but probably not, since everything else was normal. That was not the sort of reassurance Kristen had hoped for; she called Clayton in tears.

Fortunately, the neurology exam revealed no issues. But Cara's eyes continued to engage in optical acrobatics. The Greenes went to see a friend's father who is a pediatric ophthalmologist. He was concerned enough to send them to Duke Eye Center, at Duke University, to see a pediatric neuroophthalmologist, who scheduled an MRI for the day before Thanksgiving.

The MRI came back normal. "It was a week of celebrating," says Kristen. But the festivities were short-lived. Cara's eyes continued their dancing, and her fingers began to tremble, making it hard for her to pick up food from her high-chair tray.

The Greenes brought Cara back to Duke, where pediatric neurologists suspected an autoimmune disorder sparked by the fever that had developed before her eyes began to spasm. They started treating her with steroids and intravenous immunoglobulin. Then in January 2014, when a test revealed that Cara's retinas weren't working as they should, specialists wondered if the cause just might be genetic. Cara's symptoms didn't seem to fit a classic autoimmune pattern, so Shashi, the geneticist, was summoned by pediatric neurologists to take a look.

Shashi still has the lyrical lilt of her native India, where she practiced pediatrics and saw many children with chronic health problems related to genetic disorders. That experience inspired her to seek out a genetics fellowship at the University of Virginia. Shashi has dark hair, thick and curly, that she wears loose, a broad smile, and boundless patience and time for the children she cares for. "These families go through a lot," says Shashi. "For at least half of these patients, just getting a diagnosis is a challenge. So many of these problems are not treatable. You just manage them."

On the steamy August morning I spent shadowing her, I noticed that she spends far more time per patient than the typical physician. Shashi's clinic visits average ninety minutes with each patient, an unheard-of luxury in contemporary health care. Both before and after an appointment, she devotes several hours to researching a patient's condition and to coordinating care—physical therapy, occupational therapy, social work—to help improve quality of life. In so many cases, Shashi has no magic potion, no cure or treatment, to offer her patients. All she can do is help make them more comfortable. "For every patient we see in clinic, we spend four to five hours but get reimbursed for just an hour," says Shashi. "It's not a financially lucrative model. Geneticists are money-losers for medical institutions, but we bring a valuable service to the patients."

It's a sobering financial irony: genetics—how our heredity affects our well-being and how our genes interact with our environment—is increasingly becoming a focus of health care. Yet the doctors who specialize in this field are practically considered a financial liability.

But they are a psychological stronghold. In the past few years, they've been able to offer patients and their families more hope, more truth, than they've ever had before. Before the advent of sequencing, half of children with presumed genetic disorders never got a diagnosis. "When I first used whole exome sequencing, I kind of felt like I was on the horizon of a new frontier," says Shashi. "I could look a parent in the eye and say we have a technology available to you that we have never had in the past. To be able to do that was just amazing. I have been in genetics long enough that I know the sense of defeat when we see these intractable patients year after year, so this was invigorating."

Diagnosing rare disease is imprecise because gene discovery is unfinished business. There are more than 7,000 single-gene disorders, but we know the genetic basis underlying only about half of them. Not infrequently, the evidence that a particular gene causes a specific disorder is woefully slim—reported in just a family or two in the medical literature. Drawing these connections, one by one, between genes and disease is a painstakingly slow process. The good news is that as more families are identified with the same symptoms and mutations in the same genes, the evidence will be strengthened, the connections fortified.

Rarely is diagnosing a genetic disease as straightforward as identifying, say, an ear infection. When a pediatrician wields an otoscope to peer into a child's ear canal, redness and inflammation are easily seen. They are the calling cards of otitis media. But genetic disease is trickier to suss out. Yes, geneticists are trained medical practitioners. But in their day-to-day work, they may more accurately be described as detectives.

Their sleuthing increasingly centers on sequencing, which requires just a teaspoon or two of blood, nothing more than a routine blood draw. Then reagents, substances that wash and purify the DNA, are added. The end result most closely resembles a ball of snot. That stringy blob is DNA, which is extracted into tiny tubes, then fed into a sequencing machine. That's the easy part; the hard part is analyzing the data.

Software can do the initial heavy lifting, highlighting potential variants, or genetic changes, of interest. But computers can only do so much; once a list of possible culprits has been drawn up, it's up to a human brain to do the rest. In their quest for a diagnosis, clinicians pore over their data looking for "candidate genes" that might be responsible for a patient's symptoms. They look for specific variants that have previously been associated with a condition, or, more broadly, for variants in a gene that has been linked to disease. They dig through medical journals, on the look-out for other patients with similar features of disease. They email colleagues across the globe, listen closely at medical conferences for look-alike cases, and scour the Internet for mention of a child with the same symptoms.

Sometimes they cast the net wider by cross-referencing their data, comparing mutations that appear to cause loss of function in a gene in a sick child with the population at large to see how often the mutation occurs in healthy people. If a gene change is extremely rare, researchers have more confidence that it's fueling a child's illness. Often, animal models are created with the same mutation as a human to see if a sick child's symptoms are reflected in an animal "knockout"—so called because the function of the gene in question has been knocked out, or silenced. This proved helpful at the University of Washington (UW), where two families were identified with a mutation in the *MYLPF* gene. One child had a severe club foot that wouldn't respond to surgical correction and had to be amputated. Upon dissection, the foot turned out to have no muscle, which is why it couldn't be remolded: there was no muscle to hold the bone in place. When a mouse knockout of the *MYLPF* gene was observed to be lacking muscle, it was the proof that scientists needed to link the gene with the child's condition.

In 2015, Mike Bamshad was the senior author on a paper that laid out the incredible progress researchers have made in connecting the dots between specific genes and specific diseases—and how much work remains to be done. Diseases that result from a change in just one gene are classified as "Mendelian," after Gregor Mendel, who identified the basic principles of heredity. These diseases are said to have Mendelian phenotypes. A phenotype is a person's physical characteristics; in regard to illness, a phenotype is essentially each person's individual symptoms of disease.

The paper, published in *The American Journal of Human Genetics*, detailed a worldwide effort that enlisted 529 researchers at 261 institutions in thirty-six countries who examined 18,863 samples from 8,838 families. Their investigation yielded 579 known and 470 novel Mendelian phenotypes as of January 2015. All told, this extensive collaboration between researchers around the world and the U.S.-funded Centers for Mendelian Genomics (CMG) has pinpointed 956 genes, including 375 that hadn't yet been associated with disease, that correlate with one Mendelian condition or another. Some of the more well-known Mendelian disorders include cystic fibrosis, Huntington's disease, and muscular dystrophy.

It sounds like researchers have accomplished a lot, and they have. But what's frustrating is how much more we have left to learn—what the paper's authors call a "formidable gap in biomedical knowledge."

The Centers for Mendelian Genomics—there are five, including at UW, where Bamshad is a co-leader—are working on that. Established by the National Human Genome Research Institute and the National Heart, Lung, and Blood Institute (NHLBI) in 2011, the CMG sites are laser-focused on pinpointing the areas in the genome that contribute to Mendelian disease. They rely on whole exome sequencing, analyzing the protein-coding part of the genome, to home in on new genetic variants responsible for these rare single-gene disorders.

Starting in 2016, CMG's counterpart, the Centers for Common Disease Genomics (CCDG), is using sequencing to examine the role that genes play in common diseases such as diabetes, stroke, heart disease, and autism. CCDG researchers expect to sequence hundreds of thousands of genomes of people with and without these diseases to compare how genomic variations between people influence disease risk.

The ability to diagnose rare disease isn't new. Doctors have been sharing information about their mystifying patients who have defied diagnosis for years, through casual conversation in the hallway, presentations at academic meetings, and published reports in medical journals. Now, sequencing technology enables researchers to scan thousands of genes at once for a causative culprit, and

Internet technology enables families of children who share a genetic mutation to connect. Bamshad calls this latter phenomenon "social networking for gene discovery." In 2016, he and colleagues launched MyGene2, a website where families of kids with rare disorders, both diagnosed and undiagnosed, can post their children's medical data and personal stories. The more families that share data, the better the chance of connecting genetic mutations with disease.

❏

On the day that Shashi was called to examine Cara, she, like her colleagues, emerged from the examination bemused. Cara's symptoms didn't point to any genetic disease in particular. It seemed like an opportunity to marshal the substantial genetic firepower of sequencing, so she told Cara's parents about the test that could scan all of their daughter's protein-coding genes—the exome—at once. The Greenes doubted that genes were to blame; after all, they were perfectly healthy, and no one in their extended family had experienced Cara's symptoms. It seemed more plausible to them that Cara had an autoimmune disorder. It was easier to wrap their heads around that sort of diagnosis. They declined Shashi's offer, overwhelmed by its potential implications.

The Greenes desperately wanted to believe that the autoimmune treatments—steroids to decrease inflammation and intravenous immunoglobulin, purified antibodies thought to halt an autoimmune attack—were working, and it appeared, to an extent, that they were. Some days, it seemed that Cara's eye spasms had eased. But on other days, her eyes flitted up and down, back and forth. And recently she'd begun to experience odd new symptoms, having trouble lifting her arms to shield her tiny body from a fall. For a brand-new walker, this presented a formidable challenge. Cara became reluctant to take a step.

At a loss for what else to do, Cara's neurologists had recommended chemotherapy to zap her immune system, which would hopefully trigger some sort of internal reset button and put an end to her symptoms. It wasn't a treatment to be undertaken lightly: the chemo would come with nasty side effects; it was, after all, toxic medication. The drugs were scheduled to start dripping into Cara's veins the week of April 20, 2014.

But in the meantime, almost as an aside, the Greenes had grudgingly acquiesced to Shashi's repeated requests to sequence Cara's exome. "They really did not want a genetic diagnosis, because they thought that a genetic diagnosis was untreatable," says Kelly Schoch, a genetic counselor who works with Shashi.

"That's right," recalls Shashi. "They thought genetic things were bad and had no treatment."

"Which is often true!" says Schoch.

Often, but not always.

The most baffling diagnostic challenges may wind up at one of seven centers around the country that are part of the National Institutes of Health's Undiagnosed Diseases Network. An expansion of a related program established in 2008 and headquartered at the NIH, the network has the goal of providing an interdisciplinary approach to cracking the mysteries of rare disease. Evolving genomic tools are at the center of this campaign. With costs dropping for sequencing and with testing improving, in 2014 the NIH expanded its program from its clinic in the Washington, D.C., area to six additional sites around the United States, including Duke University, where Shashi is a co-leader. Baylor College of Medicine in Houston, Stanford University in Silicon Valley, UCLA, Vanderbilt University in Nashville, and a consortium of three Boston hospitals affiliated with Harvard Medical School round out the network, which relies heavily on genomic approaches such as sequencing to identify disease. Each of the centers is expected to tackle 50 new cases a year, in addition to the 130 patients worked up annually at the program's hub.

Doctors who are stumped by symptoms can apply to the network on behalf of a patient. There aren't hard-and-fast criteria for who will be accepted, but a patient must have first undergone a thorough workup that led nowhere. If multiple family members are affected, that increases the chance that one of the centers will take on the patient's case. At Duke, patients are funneled into the program by various referring physicians and Shashi, whose unassuming demeanor works well when wooing potential patients. It doesn't hurt that many of these patients are at their wits' end.

The NIH's main clinical center has cracked just 25 percent to 50 percent of its cases. At Duke, where Shashi has been engaged in medical sleuthing since 2010, the overall resolution rate is 38 percent. Solve rates remain relatively low because so many disorders have yet to be matched with genes. And, of course, some patients turn out not to have a single-gene disorder at all. There are whole categories of genetic disorders that sequencing can't detect.

Epigenetic disorders, for example, are caused by events outside our genome that influence the way that information inside our genome can be used; the information is still there but it's off-limits to our cells. Essentially, DNA sequencing is unable to foretell epigenetic changes because it can't predict when genes will be turned on or off, a phenomenon that's known as gene expression. In some cases, a gene responsible for making a protein or protecting against disease may be turned off, rendering it invisible.

As technologies continue to improve—and the network of centers continues to collaborate—Shashi anticipates that rates of diagnosis will rise. In her initial research, published in 2012, that looked at how effective sequencing might be at helping to solve undiagnosed disease cases, Shashi and Goldstein enrolled twelve patients who had different symptoms that appeared to be caused by malfunctioning genes. She anticipated being able to diagnose 20 percent of them via sequencing, but the test ended up being even more useful than she'd anticipated. "We solved two right away, and within a year, we had solved nine of the twelve," she says. Geneticists on the undiagnosed disease beat are adept at perseverance. "We never give up," says Shashi.

The study itself was the first to take a group of kids with very different, unrelated conditions and try to use sequencing to understand what was making them sick. More commonly, researchers scrutinize a group of patients with similar symptoms and scan their genomes for the culprit variant. Focusing on children with disparate symptoms was an ambitious project—and it came about pretty much by accident.

In March 2010, Shashi and Goldstein met each other in Washington, D.C., at a National Institutes of Health meeting (even though they both worked at Duke, they didn't know each other). They began chatting, only to realize that they were booked on the same flight back to Durham; when it was canceled, Goldstein asked Shashi if she wanted to rent a car and drive the five hours home. "That was serendipity," says Shashi. During the trip, they talked about their mutual interests. He told her that he was a human geneticist who does basic science research. More specifically, he oversaw a lab that was using sequencing techniques to try to get to the root of illness. She told him that she was a clinician who sees patients whose diagnostic odysseys he could help resolve. "By the time we reached Durham," says Shashi, "we'd decided, 'Let's do this.'"

On March 12, Goldstein replied to an email from Shashi about setting up a time to meet, writing that he thought the study "could really be something very significant." He was right.

The study's success prompted Shashi and Goldstein, in 2012, to start an undiagnosed-disease clinic. They began seeing patients one day a month and have increased to two days, for a total of eight patients a month. (Goldstein moved in 2015 to Columbia University to launch its new Institute for Genomic Medicine.) To start the investigation, Shashi sequences trios—not just the child but also her parents—to narrow down the possible genetic culprits. "We have so many variants that come up on first pass," she says. "Once you compare the sequence results of a child with the parents, you see that many of these variants are inherited from the parents, so you can put those aside." (In other words, if a child shares a variant, or genetic alteration, with a parent and the parent has no symptoms, odds are that the variant is meaningless.) The trio approach is what Shashi used with Cara Greene.

☐

When Shashi first saw Cara in early 2014, she observed the tiny brunette's jerky eyes and unsteadiness on her feet. "Her arms were completely flaccid, she couldn't move her shoulders at all, and her forearms were weak," recalls Shashi. "Her wrists had very little movement. She had just started to have trouble swallowing. She would do a hard swallow where we heard her swallow to get the food down."

Shashi suspected a genetic problem, but persuading Kristen and Clayton to sequence Cara still proved challenging. They thought the steroids and immunoglobulin that Cara was taking were producing modest improvements. Shashi disagreed. But all she could do was wait. Several months passed and so did Kristen and Clayton's optimism. They reluctantly agreed to get tested along with Cara.

Shashi was thrilled. So confident was she that Cara had a genetic condition that Shashi asked Goldstein to fast-track Cara's testing. It took three weeks. There was no time to waste: Cara was due to start chemotherapy.

Research studies proceed according to strict protocol. All aspects of a study are scrutinized by an institutional review board, or IRB, experts who serve on a committee that oversees and vets studies being conducted at research universities to make sure they meet federal, institutional, and ethical guidelines. In order to protect patients from potentially inaccurate conclusions, doctors can't treat based on research results; regulatory guidelines require that the outcome must first be confirmed in a clinical lab.

As technology has evolved, the price to sequence a human genome has dropped precipitously, plummeting from $17.5 million in January 2005 to $47,000 in January 2010 to the relatively paltry sum of just under $4,000 in January 2015. But those price tags don't account for the arduous work of interpreting the data and determining which variants may be disease-causing. As a result, cost continues to be a major consideration when geneticists feel a patient could benefit from the test. Few, if any, insurance providers are quick to cover the testing, effectively shutting down access to this sophisticated analysis to patients who can't afford to pay out of pocket. At some institutions, such as Duke, the cost is absorbed by the hospital when insurers refuse to pay. "Access is definitely an issue," says Duke's Shashi. "A child in rural North Carolina who can't make it to a center like ours, they don't get sequenced."

At Goldstein's lab in Durham, Cara's blood was being whirled, spun, and analyzed. At the Greenes' home in Raleigh, half an hour away from the million-dollar machines that were trying to coax a genetic narrative from Cara's blood, Cara was taking a nap. It was 11:15 a.m. on Friday, April 18.

Two days earlier, Kristen had emailed Schoch for an update. "Schoch said on the first pass, they didn't see anything," says Kristen. "And we were relieved because in our heads we thought genetic meant you can't do anything about it. She said she would let us know for sure after they go back through. We were thinking, based on what they'd told us, that we should go ahead with the chemo because it was an autoimmune disorder. Monday was going to be 'go time.'"

The Greenes were growing increasingly more despairing. Cara had continued to get worse. Her arms were so weak she could barely lift them. She couldn't use her regular hot-pink high chair because the tray was too high; her parents switched her to a model with a lower tray. She seemed weaker by the day and no longer had energy to play. "It was really hard," says Kristen. "Clayton and I took turns hyperventilating. We were terrified as we saw her lose abilities. I would put toys in front of her and she would say no. She couldn't play with them. But we were confident we were on track with the autoimmune disorder and we hoped that things would get better eventually. We both have a lot of faith in God. It wasn't that we believed He was going to heal Cara, but we were able to reevaluate what we wanted for our life and for our children. Even if you are a parent who says I don't care if my child fits into the American dream, I think everyone wants their child to be smart, grow up, be good at sports. It forced us to step back and ask what we believe about the world and what do we want for our family. We realized that nothing is promised to you in this life."

As Cara snoozed in her crib that Friday morning, the phone rang. Shashi was on the line. She had some surprising news—but even more surprising was that she couldn't tell Kristen because of the research nature of the study in which Cara was participating. But Shashi could tell Kristen that the news was significant enough that she had gone ahead and petitioned the IRB committee to bend the rules and allow her to share her suspicion about what was going on inside Cara's body.

Cara was doing so poorly at that point that Kristen and Clayton assumed the worst. They figured that Shashi was going to gently inform them that their daughter was going to die. "That weekend we cried a lot," says Kristen, who was nearly eight months pregnant at the time with Cara's little sister. What if whatever genetic problem Cara had was mirrored in her unborn sibling? She passed the weekend in front of the computer, searching for conditions that matched Cara's symptoms and sobbing. On Saturday, she, Clayton, and Cara went to church. "They had a little Easter egg hunt and all the kids were running around playing with eggs and she couldn't play. So I stopped at Food Lion to get an Easter basket because I thought, What if she's not here next year?"

On Monday morning, Shashi called again. At 1:30 p.m., the Greenes were scheduled to meet with neurology to discuss details of the impending chemotherapy. Shashi asked if they could show up at 1:00.

Aware of the family's plight, Goldstein's lab had expedited the data-crunching. The week before Cara was slated to start chemo, Shashi had received an email that a researcher in Goldstein's

lab had winnowed the list of potential genetic troublemakers down
to two variants, then compared those variants with a Duke database of 1,200 people's genomes to see how unusual they were.

"One thing didn't make sense," Shashi told me. That variant
was associated with people who have Hodgkin's disease, and didn't
explain Cara's symptoms. In the other gene, *SLC52A2*, the researcher had found compound heterozygous mutations—two mutations, each different, both within the same gene. One variant came
from Cara's mother, the other from her father. This seemed more
promising.

First, Shashi looked at *SLC52A2*, the gene that contained the
two variants, to see if it had been linked with a particular disease.
Next, she looked at both variants to see if they were known troublemakers. Then she went deeper, examining the nature of the
gene changes themselves. Three DNA letters together make up
what's known as a triplet code for an amino acid. If even one letter
is missing or erroneous, the amino acid is not being synthesized
correctly, which impacts the protein it manufactures and can result in disease. "Sometimes you're not sure," says Shashi. "If one
letter is substituted for another, for example, you can use computer models to predict if it will be damaging. No one would
hang their hat on it, but it's one of the pieces of information you
use to try to assess a variant." The more damaging the variant, the
more likely that a researcher will give it a closer look.

"For Cara, because I was so focused, it took me about four
hours to figure things out," she says. "For some, when the phenotype is not specific and there's no one hallmark feature that stands
out, it could take months. Sometimes we just have to say that we
have a candidate gene and leave it at that."

In researching the variants in Cara's *SLC52A2* gene, Shashi
came across a description that sounded just like Cara. "When I
looked in the literature, it was clear," says Shashi. "Cara had Brown–
Vialetto–Van Laere syndrome." Brown–Vialetto–Van Laere syndrome was discovered in 1894, and fewer than a hundred people

worldwide had been diagnosed. "Cara's phenotype matched everything to a T. Her clinical progression, her symptoms, the inheritance background, and the two damaging variants clinched it for me."

Typically, after Shashi settles on a diagnosis, she sends that patient's DNA to a clinical lab that has been certified by the federal government. According to Duke's IRB protocol, a lab can't take results obtained through research and use them for patient care unless those results are first confirmed by an outside clinical lab that's part of the highly regulated Clinical Laboratory Improvement Amendments of 1988, an update to an earlier law that established oversight of clinical laboratory testing. In most cases, the six weeks or so that clinical confirmation adds is not a problem. But in Cara's situation, it could mean that her entire immune system would be wiped out with chemotherapeutics.

On Monday morning, Shashi and Goldstein spoke with the IRB committee chair and explained the situation. Sophisticated genomic analysis had indicated that Cara's deteriorating body could be strengthened with a simple vitamin. Chemo was unnecessary, irrelevant to Cara's condition. The committee chair agreed and gave Shashi the green light to inform the Greenes.

Kristen was jittery with anxiety while she and Clayton waited at the hospital with Cara. When they entered the exam room, Shashi cut to the chase: "There is a treatment option and we are going to start it today," she said and held up a piece of paper with BROWN–VIALETTO–VAN LAERE SYNDROME written in all caps. "This is what Cara has," said Shashi, "and this is how you pronounce it."

Brown–Vialetto–Van Laere syndrome—BVVL, for short—is a rare genetic disorder characterized by nerve damage. It can cause upper limb weakness without lower limb weakness, which is exactly how it presented in Cara. Cara's condition is caused by a problem with a gene that codes for a protein that transports riboflavin into cells. The error keeps riboflavin from embarking on its cellular journey, which means that it's not getting the chance to convert into two cofactors that produce a chain of chemical reactions that makes energy. That energy is used by cells in the process of metabolism. Simply put, too little riboflavin results in too little energy.

Cara has a metabolic disorder, a serious imbalance of energy production. Two of the research papers that Shashi had read when researching BVVL had discussed treatment with riboflavin for children with the condition. Riboflavin is part of the vitamin B family, vitamin B2 to be exact. Cara needed a B2 boost. No powerful drugs and no chemotherapy required.

"We are going to treat her with a vitamin," Shashi told Cara's parents.

Clayton put his head down on his knees. Kristen laughed. They were incredulous. "A vitamin?" said Kristen. "Are you serious? She just needs more of a vitamin?"

It certainly seemed that way. "We understand this gene more than some others," Shashi told me. "Maybe we just need to give larger doses of riboflavin and some will enter the cell. It was easy to say let's try it."

Cara now takes mega-doses of riboflavin, an orangey-yellow substance that is naturally found in milk, cheese, leafy vegetables, and mushrooms. Cara needs far more than what's bioavailable in food. She takes 800 milligrams per day, in contrast to the normal requirement of 1 to 2 milligrams for a healthy child. That enormous dose is enough to turn her urine the color of an orange Fanta soda. In the summer of 2014, shortly after the Greenes moved east to Wilmington, I visited them at home. Cara followed Kristen and me around, banging plastic toys together. When we entered her room, Kristen pointed out Cara's crib. "That's Cara's bed with the riboflavin stain on it," she says. Does it go without saying that orange is the family's new favorite color?

Cara's diagnosis had come not a minute too soon; what Cara's parents didn't know at the time was that she wouldn't have survived had sequencing not deciphered her diagnosis.

"This is a progressive condition, and it's typically fatal," says

Shashi. "She could have gotten a bad chest infection and with her diaphragm so weak, she could have aspirated. She would have died. This test saved her life."

Cara's situation is, in fact, a textbook example of how sequencing can revolutionize health care. Sequencing revealed a diagnosis, and her diagnosis was paired with an extremely effective and nontoxic treatment. But as inspiring as Cara's story is—imagine the headline: Girl Averts Death by Taking Vitamin—it's not the norm.

"Cara's case is an outlier," says Goldstein bluntly. "Cara's result is in many ways as good as it gets. We find out what's happening with her and it results in transformational change. This is rare, but it does happen. If this is all we had, we'd still do the work because once in a while, we can truly make a difference."

When I met Cara in August 2014 at Duke, she was four months into her miracle vitamin treatment. She had returned to Duke for a check-up with Shashi and Goldstein. Cara's stubby brown ponytail was cinched with a yellow ribbon. She was wearing pink shorts and a T-shirt with popsicles, a fitting choice on this oppressively hot summer day, and she warmed to the audience crammed into the exam room: Shashi, Goldstein, and the genetic counselor Kelly Schoch; her dad, Clayton; her mom, Kristen, who was holding Cara's baby sister, Susan. And me.

Cara explored the waiting room with me while her parents spoke with Shashi and Goldstein. We padded down the tiled floors together. When she started to sway like a child who has just twirled round and round to the point of vertigo, I scooped her up. In the beloved children's book *Knuffle Bunny*, Trixie, the protagonist, throws a tantrum after her bedraggled but adored stuffed rabbit is lost. "Trixie bawled," Mo Willems writes. "She went boneless." Any parent knows viscerally what it feels like to pick up a limp, heavy, "boneless" child. That is what Cara felt like in my arms—gorgeously baby-soft but amorphous, as if her skeleton were made of cooked linguini.

When she squirmed down and wanted to walk, I held her

hand firmly to avert any tumble. She wobbled like a child just learning to walk, and at one point in the exam room, she swayed back and forth as if she were drunk. That was an improvement over her previous visit. "She's walking so well!" says Schoch.

"It's really remarkable," agrees Shashi.

Cara was already able to raise her arms more than she could a few weeks earlier. Instinctively, a child—or an adult—about to hit the ground will thrust her arms out in a bid to protect her face and soften the impact. But when Cara fell, she'd land flat on her face, because her arms had sustained nerve damage. Nerves heal slowly, at the rate of a millimeter a day. Shashi expects Cara to continue to heal, though whether she'll regain all her strength is uncertain, because there's not a huge cohort of kids who have BVVL to look to for guidance. Nonetheless, Cara has continued to improve. As of the summer of 2016, she could fully lift her arms above her head and use her hands to hold a spoon or a pencil. She can color and draw. And she can now discern colors, which is surprising because previous testing had revealed damage to retinal cells that play a role in color perception.

"Even the skeptic in me has to say this is very good," says Shashi. "Cautious as I am, I am really, really encouraged."

Kristen and Clayton, grateful for every inch of progress, are reveling in their newfound normalcy. Recently, they were eating at a restaurant and had to push the food trays away from Cara. "I told Clayton, this is what it's like to eat with a toddler who has arms that work!"

Goldstein and Shashi estimate that forty people in the United States have BVVL, although Cara is the only one with a definitive diagnosis. It gives new meaning to the concept of a one-in-a-million child. When the Greenes asked Goldstein how unusual it is that two carriers of BVVL would meet and marry, he responded: "Well, it's very, very unlikely that you would have met, but it had to happen to someone."

Could the Greenes have learned about Cara's condition before

she was born? When she was pregnant, Kristen was offered genetic testing, but she turned it down. She's a self-professed information nut, but she's also a big worrier. With no family history of
genetic disease, Kristen had assumed she was in the clear. "I thought
that this kind of stuff doesn't happen to me," she says. "I am embarrassed about how naïve I was."

Kristen's decision to forgo prenatal testing didn't make a difference. BVVL is so rare that standard carrier screening wouldn't
have tested for it. "The only way we would have known about
BVVL is if every baby had their exome sequenced," says Kristen,
her tone indicating that there's no way this would ever come to
pass. In fact, I told her, a major research study is examining the
feasibility of sequencing newborns. "Sequencing should be able to
find this," says Shashi. Detecting the disease at birth, before symptoms appeared, would have spared the Greenes heartbreak, time,
and money. Most important, it would have spared their daughter
from getting sick in the first place.

8

The Genie in the Bottle

Sequencing Newborn Babies

In 2010 in Texas, Jennifer Garcia had a baby, a little brother for her four-year-old son. She named him Cameron. Garcia had opted to do prenatal testing for conditions including Down syndrome and cystic fibrosis with both boys. The tests came back fine. Once her sons were born, she didn't think twice about having their heels pricked in the hospital and the resulting droplets of blood scanned for about thirty diseases that make up the standard newborn screening test administered to babies born in hospitals throughout the Lone Star State.

Months passed and Cameron grew, lifted his head, smiled at his parents. He looked healthy and strong, hovering in the 90th percentile for height and weight for babies his age. He laughed at the family dog. He learned to log-roll across a room to reach a toy. Then, at seven months old, he got pneumonia. In the hospital, he suffered seizures and had to be intubated. CT scans and MRIs followed, then EEGs, spinal taps, and blood transfusions. No one knew what was wrong. First, doctors thought Cameron had meningitis, then pertussis, then tuberculosis, so they plied him, just in case, with anti-seizure medications, antibacterials, antivirals, and antifungals. Specialists came and went, teams from critical care, pediatrics, neurology, epileptology, toxicology, immunology, infectious disease, respiratory therapy. Ten days after he was admitted to a major medical center in Houston, an answer to what was ailing Cameron finally emerged: an immunologist suspected Cameron had severe combined immunodeficiency, a genetic disorder otherwise known as "bubble boy disease." Children with severe combined immunodeficiency, or SCID, don't have a functioning immune system, which was why Cameron wasn't getting better.

The diagnosis perplexed Garcia and her husband, John. They had no family history of SCID. In fact, they'd never even heard of it. In any case, wasn't Cameron's newborn screening test supposed to pick it up? Garcia started researching, and what she found left her in disbelief. SCID is detectable via newborn screening, using the same dried blood spots that the Texas Department of State Health Services analyzes for the other diseases for which it scans. But Texas, along with most states at the time, didn't screen for SCID. When SCID is identified early, before a baby falls seriously ill, a bone marrow transplant usually can cure the otherwise fatal condition, as it serves to replace the compromised immune system with a healthy version. More than 90 percent of babies who receive transplants in the first three and a half months of life recover. Cameron was already eight months old at his diagnosis, desperately ill and fighting for his life.

Cameron Garcia was born just one month after SCID had been added to the national list of recommended core newborn screening conditions. Yet more than two years would pass before Texas would begin screening every baby for SCID. That was far too late for Cameron, who died on March 30, 2011. He was nine months old.

Understandably, Cameron's mother, Jennifer, emphasizes the downsides of not screening for a disease if it's technically feasible. Since the night she left the hospital without Cameron in her arms, Garcia has become an activist who was ultimately instrumental in persuading Texas to include SCID among the diseases for which it screens. Knowing that all babies born in Texas hospitals are now tested for SCID makes Garcia's loss marginally bearable. "I wanted people to know this little baby changed things and opened eyes for a lot of people," Garcia said in a video about the importance of screening for SCID. "If we would have known Cameron had SCID, and we could have found that out earlier before he had any infections, I'm absolutely one hundred percent certain that Cameron would be here today. I wanted his little life to have meant something not just to our family."

But what if we didn't have to go through the time-consuming process of adding new diseases, one by one, to the list of disorders that newborn screening can detect? What if one test could look for many of the diseases that newborn screening identifies, plus lots more?

The question is not hypothetical. In highly anticipated research that stands to overhaul what we know about health from the first moments of life, the National Institutes of Health has charged four university medical centers with studying the medical, behavioral, economic, and ethical implications of using genome sequencing to map out the entirety of babies' genetic code. Would it be wise to sequence every baby's genome?

There are obvious benefits. Far more children who are at risk could be identified, allowing earlier treatment for someone whose life, like Cameron Garcia's, hinges on early detection. But inevitably, some parents will have to cope with finding out about health problems that can't be mitigated, and about the genetic missteps called "variants of uncertain significance" whose impact is unclear: they could indicate a problem or they could simply be a string of DNA gobbledygook.

Depending on what results are returned to parents, many moms and dads will wind up finding out that the bulk of their child's genome is still incomprehensible. Michelle Huckaby Lewis, a trained pediatrician and lawyer who researches genetics policies at Johns

Hopkins's Berman Institute of Bioethics, worries that could cause problems. "The genetics and subspecialty workforces will not be staffed adequately to meet the growing demand," she wrote in a commentary in the American Medical Association's pediatrics journal. "Moreover, coveted appointments with subspecialists may be filled by children whose conditions may not manifest until later in life, making access more difficult for those whose needs are more urgent."

Regardless, it seems to be the direction in which health care is headed. "We are moving to a world where the technology will get so good and the cost will get so low that it will be very appealing to apply sequencing to not only sick people but well people," says the geneticist Robert Green. Green co-leads the BabySeq Project, the newborn screening study at Harvard-affiliated Brigham and Women's Hospital and Boston Children's Hospital, one of the four federally funded sites. BabySeq is examining how parents and doctors can use genomic data to improve children's health care. Green and his co-leader, Alan Beggs, who helped sponsor the contest to diagnose Adam Foye, are studying 240 sick and 240 healthy newborns. They are randomly sequencing half of each group, to assess whether parents of sick kids respond differently to sequencing results than parents of healthy babies. Do parents of sick babies find the additional information helpful while parents of babies deemed healthy find it overwhelming? Does either group prefer the more limited picture provided by conventional newborn screening? What's the best way for doctors to incorporate this wealth of data into caring for the youngest and most vulnerable patients? The intent, says Green, is to explore some key questions: "Is this scary or not, is this useful, is this likely to confuse the hell out of people or not?"

In a lead-up to the study, Green and colleagues surveyed parents soon after their child's birth to ask if they'd want to sequence their baby's DNA. They found a groundswell of interest in newborn sequencing. Three months later, they went into greater detail, explaining to parents exactly what kinds of data that genome sequencing could generate about their children—cancer risk, for example, or predisposition for Parkinson's disease.

The percentage of parents who remained interested hardly budged. "This suggests there is a gigantic appetite out there for this, even in healthy babies," says Green. "It is going to be hard to resist."

Still, sequencing a baby and "vomiting the results out to the family," as Green characterizes it, "feels like it's very dangerous." The combination of anxious parents and doctors trying to interpret uncertain results seems particularly volatile. "People are a bit more sanguine about finding out stuff about themselves than they are about their kids," says Green. "The salient question is harm. Depending upon who you talk to, there are all these theories about harm—about anxiety, distress, misconstruing information. All these questions are heightened when talking about babies because they aren't able to have a choice. This is a first opportunity to look for harm."

When I visited Boston in the spring of 2015, the project was on the cusp of recruiting its first infant. I thought I'd meet with a researcher or two, but I was greeted by half a dozen— neonatologists, geneticists, and genetic counselors. It takes a village to raise a child—and to hash out the details of sequencing that child. They explained that BabySeq (which, by the end of 2016, had enrolled about 100 families) would limit the results it returns to parents to only those gene changes that are linked to diseases that take root in childhood. The infants' parents and their pediatricians will also be enrolled in the study, with the goal of assessing medical outcomes and impact on parent-child bonding, as well as whether the data is useful and how it is incorporated into a child's health care. In other words, does the massive influx of information from genome sequencing translate into better health care for a child? Does the benefit justify the costs, financially and emotionally? "If you imagine a world where every baby could be sequenced quickly, how would that information be used by their doctors to facilitate their care, to make a diagnosis, to prescribe medication?" says Green. "We're trying to model that situation at a time when it's not really easy or cheap to sequence and doctors aren't used to dealing with it. We're trying to model the future."

But not a speculative, far-off future, if Green's predictions are correct. "In five years, I am suggesting that sequencing will be given away as a freebie," he says. "Your bank will give it away for opening an account. You can trade in your frequent flyer miles for it. Your gym will be giving it away."

There are certainly upsides.

But it's difficult to predict on a case-by-case basis how a specific family will process unsettling news. Take the genome-sequencing contest that diagnosed the cause of Adam Foye's muscle weakness. While it ended years of uncertainty for his family by diagnosing the middle-schooler with a titin (*TTN*) mutation, Beggs didn't know how Adam's parents, Sarah and Pat, would react. Not only did they learn that they each had a similar change in their *TTN* gene, but they also learned that the genetic variant was associated with a heart-related condition called cardiomyopathy. Further testing revealed that Sarah Foye, her mother, and especially her sister had developed cardiomyopathies. In other words, in learning what had gone awry in Adam's body, they unexpectedly discovered something awry in their own bodies. They're now on medication to control the condition. "On one hand," says Beggs, "I can imagine this would send them off the deep end. Think about a cancer diagnosis: Some people become activists and engaged. Others become depressed. That is the dilemma in all this. You can't protect everyone from having an adverse reaction."

It's not surprising that the atmosphere is particularly charged when it comes to newborns, who are the equivalent of a blank page. Their future feels wide open; is it wise to disabuse parents of this belief?

In actuality, no one's future is truly wide open; everyone is made up of genes, and genes contain the imprint of generations past. This is what's unique about the role of genetics in health care: genes offer insight not only into an individual patient but into that patient's family. Genes tell us about our relatives, and this has helped give rise to the concept of "genetic exceptionalism." Supporters of genetic exceptionalism believe that genetic test results, as opposed to other health-related data, deserve special privacy status because of the potential for discrimination against people with certain genetic traits or conditions. "Why do all of us undeniably have at least a nagging sympathy with the idea of genetic exceptionalism?" asked James Evans and Wylie Burke, the authors of a 2008 commentary in *Genetics in Medicine* that posited that genetic information is no more deserving of protection than any other medical data. "We would argue that there are two reasons: the first is that genetics is at the heart of our most profound relationships: DNA testing can confirm or deny parenthood and shed light on our ancestry. The second derives from a cultural belief that genetics largely determines who we are (despite many observations to the contrary)."

In part, Evans thinks that genetic exceptionalism stems from the field of genetics being considered distinct from mainstream medical specialties, in a rarefied class of its own. Many geneticists, for example, haven't been engaged in routine care of patients to the extent that other specialists have. As a result, many of their norms and rules have developed independently of other fields of medicine. Contrast the straightforward approach to getting an MRI—a doctor sends a patient for an MRI, no counseling required—with the multistep process that for years characterized genetic testing. It used to be standard practice to require a patient to make an initial appointment to talk about the advisability of genetic tests. That patient would then head home, muse on whether to proceed, and, if interested, make another appointment to have blood drawn. Results would be delivered at a third appointment. "This model was driven by the thought that genetic information was just so scary and foreboding that folks needed tons of counseling to handle it and understand it," says Evans, a professor of genetics and medicine at the University of North Carolina at Chapel Hill and the editor-in-chief of the journal *Genetics in Medicine*.

As more patients have genetic testing, that paradigm has fallen by the wayside, bolstered by a lack of evidence that people needed so much counseling. Now, counseling and testing typically take place at the same appointment. Evans and others have shown in studies that people are open to getting results by phone, and indeed often prefer that mode of return.

Although protocol for testing has been streamlined, genetic exceptionalism has persisted in the emphasis on the counseling that precedes genetic testing, which includes discussion of potential secondary findings that could accompany results. Yet patients referred for MRIs aren't told that their imaging also may reveal unexpected results.

What could explain the discrepancy? Well, says Evans, whose left upper arm features a tattooed rendering of a double helix that morphs into a Darwin-fish head, "genetics is special in a way that your blood count, your chest X-ray, your address, or your EKG just isn't—because in some important ways, our genome lies at the root of who we are at a fundamental level, both physically and mentally. Not that DNA is the sole driver of all of our characteristics, but the way we are 'wired' by our genomes is undeniably one important determinant of who we are at a very deep level."

Accordingly, we tend to give special protections to genetic information, particularly when it comes to privacy.

Beggs, who has a manner of speaking that makes him sound as if he's perpetually amused, thinks "genetic exceptionalism is bunko," even as he acknowledges that sequencing information, due to its predictive value and insight into the future, can be scary and breed alarm. "There are risks, but I feel the benefits outweigh them," he says. "We can't put the genie back in the bottle. It's out there and we should make it available to people who want it."

What's not bunko is the need for guides to help us navigate the increasing amount of genetic information that we can access, particularly if we start amassing troves of data about babies' genomes. "The challenge is, can we get enough well-trained people to explain the implications of these tests?" says Beggs. "We can afford to do that for 240 kids if someone calls to say, 'I'm freaking out about the results for my kid,' but can we afford that for the whole population if sequencing becomes widely available?"

Privacy and the right to an open future, discussed previously, also present a considerable challenge. Will sequencing newborns invite stigma and discrimination—even, perhaps, from Mom and Dad—depending upon babies' results?

Whether that risk is actually a reality is a large part of what the newborn sequencing studies aim to discover. Will data gathered by careful scientific design contribute to a "vulnerable child syndrome"? "If a parent is told their newborn is going to develop a particular disease, it would almost certainly have an impact on how that parent interacts with that child and how they think about that child's future," says Joy Boyer, a senior program analyst with the National Human Genome Research Institute's Ethical, Legal and Social Implications (ELSI) program. How much should the child's privacy matter?

I have discussed this dilemma with Debbie Horwitz, a friend of mine who survived breast cancer and has tested positive for a *BRCA* mutation that boosts the risk of disease. Horwitz, from Raleigh, North Carolina, remembers the overwhelming sense of worry she felt upon leaving the ultrasound appointment where she found out she was pregnant with a daughter. "I came out saying, 'Oh my God, what if she gets breast cancer?'" Her husband, Evan, didn't understand. "He calls me 'doom and gloom,'" says Horwitz. "But I said, 'I'm going to die if she gets breast cancer.'" Horwitz was determined to find a way to test her young daughter,

Jordan, but her husband was adamantly opposed. She now agrees with him. "I don't think it's fair for us to have that information now and have that heaviness in our family or in how we relate to Jordan," Horwitz says. "I think knowing whether Jordan is positive or negative would also cause a lot of tension and sadness in our family. When you think about testing a child, you have to think of the ramifications it would have on lots of family members."

❒

In fact, research on how test results affect parents and their perception of their newborns has shown that false-positive results on newborn screening panels create stress. Extrapolate that stress to sequencing babies' genomes, which is bound to yield all sorts of inconclusive findings, and it stands to reason that the broader pool of potential results may cause anxiety. Do we really need an official study to tell us this?

Susan Waisbren thinks so. Waisbren led the 2014 study cited by Robert Green, the study's senior author, in which 83 percent of new parents expressed interest in sequencing their babies' genomes. She believes it's important to quantify the risks and benefits of subjecting the most voiceless members of society—newborn babies—to comprehensive sequencing tests. Waisbren, a psychologist at Boston Children's Hospital, spearheaded another study with Green that tracked the parents who had been surveyed in their first study. She interviewed those parents of healthy babies between six and eighteen months after birth and administered a parental stress inventory and a questionnaire about parent-child bonding. The outcome, says Waisbren, was "quite interesting." She found that fathers who received concerning information about their babies were more stressed about it than mothers. Meanwhile, mothers who received concerning information about their babies reported not feeling any more stressed than other mothers. In neither case was bonding affected.

Are these findings offering us carte blanche to move ahead with sequencing every baby in the United States? Not exactly. Bonding is not black-or-white. What enhances bonding for one family might disrupt it in another. The temperament of the parents and the nature of the test results will play a significant role in determining how parents process data about their babies. Two people who receive the exact same results may react completely differently. But the knowledge is bound to have some effect, insists Susan Wolf, a lawyer who founded the University of Minnesota's Consortium on Law and Values in Health, Environment and the Life Sciences. "I am worried about how this changes parents' views of their kids," says Wolf. "How plausible is it to you that knowing your child's genome would not make a difference as you raise your child?"

Wolf is opposed to allowing parents to learn as much as they want about a child's future health. In other areas, such as reproductive health, mental health, and substance abuse, legislation protects the rights of teens to make their own health decisions. More than half the states allow teens to independently make decisions about birth control, prenatal care, and adoption. (Only two states and the District of Columbia permit minors to independently consent to abortion.)

Similarly, Wolf thinks that children should be able to weigh in on whether they want their genome cracked wide open and put on display for their parents. It's obviously impossible to ask this question of newborns, but it's achievable with regard to older children. "We need to talk to the kids," says Wolf, who started studying the return of genetic results in 2005. "You have to ask the person whose prerogatives and rights are being diminished, not the person whose prerogatives and rights are being expanded. Let's say researchers talk to lots of adolescents and most say that they want to make their own decision about testing. That is a strong argument."

When Wolf had her own genome sequenced in 2014, one of her kids cautioned her against doing it, asking why she wanted to hear potentially threatening information that she wouldn't be able to do anything about. Wolf proceeded anyway, convinced that her work would be enhanced by the experience. But the results she got back were hardly illuminating and were, in fact, "very disappointing." Four or five disease-causing mutations "seemed wrong," says Wolf; hundreds of variants of uncertain significance served only to jack up her anxiety levels. "It looked like a science not ready for prime time," says Wolf.

Some experts also worry about potential discrimination if babies are sequenced at birth. Congress appeared to foresee this possibility back in 2008, when it passed the Genetic Information Nondiscrimination Act (GINA), which bans health insurers or employers from denying insurance or jobs to people based on their genetic information. The Affordable Care Act of 2010 was predicated on access to coverage despite preexisting conditions; GINA ensured similar protections two years earlier. "There's nothing more preexisting than genes," says Congresswoman Louise Slaughter, an eighty-six-year-old microbiologist from New York State who sponsored the bill. But a closer look reveals the law's significant shortcomings: it doesn't apply to other sorts of insurance such as disability or life insurance. That's not because Slaughter didn't care about safeguarding those protections. But politics dictated that Slaughter, who has represented her district since 1987, pick and choose; she opted to champion people's livelihoods and their right to health insurance.

The law suffered a setback in 2016 when new regulations chiseled away at a key facet protecting employees from being pressured to share their genetic data or that of their families with their employer. The update allows employer-sponsored wellness programs to offer employees a hefty discount on their health insurance premiums if they share that data, essentially penalizing those who choose to keep that data private. Not long before, however, the pendulum swung in the opposite direction as the bipartisan Genetic Research Privacy Protection Act was introduced to further safeguard personal genetic information. The legislation strengthens protections for personal genetic data held by federal agencies, stipulating that the copious amounts of genetic data compiled from government-sponsored research studies, for example, must be kept confidential.

Rising concerns about privacy are understandable considering that sequencing will turn up far more results than standard newborn screening. Sequencing can reveal carrier status for disease, for example, in which a carrier is healthy but has an increased risk of passing disease on to any children. "States pride themselves on connecting parents with the appropriate resources," says Aaron Goldenberg, associate director of the Center for Genetic Research Ethics and Law at Case Western Reserve University. Goldenberg is examining how states might go about navigating the complexities of integrating sequencing into newborn screening. "They will go from connecting a few people to connecting everyone, because everyone will have some sort of positive result," he predicts.

But a positive result, as we've seen, is not necessarily cause for alarm. Most patients don't have a good understanding of what genes can and can't reveal, which points up the need for more general education. The more that news about genetics appears in mainstream media, the more people will start to understand the gene/environment connection. DNA is simply not the be-all, end-all in many instances. "We get this question all the time: 'When you do an amniocentesis, can you tell me if my child will be good at math or be gay?'" says Jennifer Malone Hoskovec, who serves as the National Society of Genetic Counselors' prenatal testing expert. "It's all about setting expectations. The technology is far ahead of being able to interpret in a way that's always meaningful to a patient. That is the conundrum we are in right now."

In any case, it's unlikely that states, which are responsible for newborn screening, would be eager to pay more for sequencing, which is far more expensive. The cost of newborn screening varies depending on which diseases states screen for, but it was estimated at $76 per infant in 2015. As we've seen, although the cost of sequencing has dropped well below $10,000 and there is frequent talk of the "$1,000 genome," reliably interpreting the data often costs far more. Insurance providers aren't eager to cover sequencing because of the potential for indeterminate results that can lead to expensive—and often needless—additional testing. (An increasing number of providers are paying for sequencing for undiagnosed illness, particularly when standard testing has failed to reveal a cause, as explored in the previous chapter. But coverage for sequencing of healthy children or adults is rare.)

If health insurance won't cover this type of test, it's bound to create inequity. Leaving aside the question of whether sequencing newborns is a good idea, it's pretty clear that parents with means will be able to pay out of pocket to sequence their babies while other moms and dads won't be able to afford to access this degree of information. As a result, sequencing could represent yet one more way in which society stratifies socioeconomic classes.

There are significant hurdles to surmount before we know if sequencing could or should be performed on all newborns. The cost—in terms of interpreting the data accurately (figuring out which results rise to the level of concern and which can be ignored) and allotting the time required to clearly transmit the information to families—will be extraordinary. As genomic understanding continues to deepen, newborns' sequencing data may need to be reanalyzed and parents updated on the new results. The sheer cost of supporting this counseling will lead to "overwhelming economic health expenditures," writes the Swiss researcher Jacques S. Beckmann in a commentary titled "Can We Afford to Sequence Every Newborn Baby's Genome?" in the journal *Human Mutation*:

> It is hard to imagine that the time spent exclusively with the parents can be reduced to less than half, and most likely, a full day, even if we sprint through such consultations . . . Consider a middle size hospital like the Lausanne University Hospital (CHUV) in Switzerland with 7,000 births

annually, or about 20 births per day. For such a setting, this translates to 10–20 counselors busy daily explaining to parents the genetic and clinical facts that they may (or may not) want to know and do not dare to ask. Imagine what would be needed if we opted to generalize this . . . providing appropriate pre- and posttest counseling to each and every tested individual and/or family. As a geneticist, I should be extremely pleased by such a scenario, as this would make medical genetics services among the largest ones in the hospital . . . Yet, at what cost for the health system?

Ingrid Holm, a pediatric geneticist who is part of the BabySeq Project, tends to think that for now, routine newborn sequencing—and sequencing of adults too, for the most part—is largely shtick. "Most people's genomes are pretty boring," says Holm. "There's just not much there." Holm is speaking from personal experience; in her line of work, it's de rigueur to get your genome sequenced. She was ambivalent about doing it—she's healthy, with no ominous family history—but she forged ahead. "Everyone else around me was doing this," she says. "I wasn't happy about it, but I kind of felt like I should. It becomes like cocktail conversation, trying to show that you're cool."

As for those parents who sign up for BabySeq to probe their infant's DNA, most are likely to be disappointed—or perhaps relieved. "They will think it's interesting to find out what their baby's genome is like, but they are going to have to have a fairly high level of education to understand this," says Holm.

❒

At UNC–Chapel Hill, another of the newborn sequencing sites, researchers are investigating whether moms and dads feel prepared to receive newborn sequencing data. When researchers interview parents about their attitudes toward sequencing babies, they are finding that they have to simplify the way they present information. For starters, they have substituted the word "change" for "mutation," which they found had confused people. In trying to give parents general background about genetics, the researchers realized they were overdoing it, muddling rather than clarifying parents' understanding. "Some people have never heard of a gene or a genetic mutation," says one of the researchers. "There is a continuum of knowledge."

Sometimes, couples are on the same page with each other, but it's not uncommon for them to disagree on what information they'd like. Some parents say that their natural tendency is to be anxious; these parents are probably better off receiving curated results. "They talked about living with years of worry with the chance that the disease may never appear," says Caroline Chandler, a public-health analyst at RTI International, a nonprofit research institute that is heading up this component of UNC's project. Yet other parents want all the results, and they don't want them sugar-coated; they seek as much information as possible so that they're better prepared.

As Eric Green, director of the National Human Genome Research Institute, explains: "If parents exquisitely want to know, I just do not think that we can stop them." In any case, he notes, sequencing results are just another instance of parents having information about a child's future—family health history is a prime example—and assessing when it's appropriate to share that information. "For me," says Green, "I am absolutely at family risk for melanoma from both sides of my family. And I am fair-skinned. On top of that, I have an eighteen-year-old son who is a redhead and is at even greater risk. I get regularly checked out. At some point, I told my son, 'Your entire life you will be at risk for cancer.' Once a year, we go together to get checked out. I will take my daughter too—she's fifteen. This is what it looks like to deliver genetic information to kids."

Some of the earliest data about the effectiveness of newborn sequencing comes from Inova Translational Medicine Institute in Falls Church, Virginia, a research association that uses genomic data to try to develop personalized health care. Inova has been recruiting trios—you'll recall that refers to a mother and a father, plus their baby—since 2011 by asking women who deliver in affiliated hospitals if they're interested in participating. The intent is bold: to sequence these babies at birth and track them throughout their lives to see how genomic data can lead to better health care. "We realize these are uncharted waters," says John Niederhuber, Inova's CEO and former director of the National Cancer Institute. "One could say, it's uncharted waters so we shouldn't do it. On the other hand, I don't know how you learn about these uncharted waters unless you do it."

Emma Warin enrolled in the study in 2012 when she was four months pregnant with her son, Garrett. Warin has worked in cardiac catheterization labs and as a medical device sales representative, so she feels comfortable in the medical world. "I don't get freaked about this stuff," she said when Garrett was two months old, safe and sound in his pistachio-green bedroom with a monkey painted above his crib. She had not yet received his results. "I think it's wonderful and feel very thankful to be part of this. It's information that should be available to everyone in the future."

Warin, busy with the swirl of motherhood, realized she hadn't received Garrett's results only when I contacted her again a year later to request an update for this book. She then checked in with Inova, where researchers are instructed to emphasize that a lack of worrisome results doesn't necessarily convey a "blanket statement of genomic health."

Warin doesn't remember hearing any medical caveats. She just remembers feeling relieved when she was told that Garrett had "none of the markers for anything on the list. She said she would check again, but that since no one had called us about anything, that no news was good news!"

In a study published in 2016 that reported on the results from sequencing nearly 1,700 infants and their parents, Inova concluded that sequencing is not at the point that it can substitute for traditional newborn screening; it can miss some things that screening finds. In the study, there were five individuals with a condition that newborn screening should have been able to detect; sequencing identified just two of them.

That said, sequencing can identify many more potential disorders than state-run newborn screening can. In the study, sequencing was also able to resolve inconclusive results from newborn screening and yielded far fewer false positives. Some of the advantages are practical, given that states don't necessarily screen for all of the same conditions. The study identified fourteen children with glucose-6-phosphate dehydrogenase deficiency, a condition that is included in newborn screening in some states, but not in Virginia. People with this condition should avoid certain foods and medications due to the possibility of red blood cell breakdown.

For the most part, no one is yet suggesting that sequencing newborns should replace newborn screening. Rather, it's a tool that could improve the baby's care. For example, learning that a baby is at greater risk for heart disease or cancer could pave the way for early monitoring. Learning that a baby harbors a rare mutation associated with a severe and potentially fatal reaction to the common anticonvulsant carbamazepine could save a life. Such an example is at the most dramatic end of the spectrum. But as our knowledge of genes and their effects continues to expand, the possibilities are intriguing. "The idea is that as we learn more and more, decisions about prevention, lifestyle, and treatment could be tailored via the baby's genome," says BabySeq's Robert Green.

While it's unlikely that sequencing would render newborn screening irrelevant, it surely makes sense to consider the technology complementary. For a baby such as Cameron Garcia, this sensitivity could have made the difference between life and death. Cameron had the common X-linked SCID gene change, which sequencing can detect. If Texas had been sequencing babies when Cameron was born, it wouldn't have mattered that SCID was not part of the state's newborn screening panel; sequencing would have flagged his disorder.

But there are many other gene changes that can also cause SCID, and they aren't all detected by sequencing, especially if they're the result of a novel mutation not previously known to be associated with disease. The TREC test, used in newborn screening and developed by the pediatric immunologist Jennifer Puck, can pick up cases of SCID regardless of the genetic cause. Puck is the lead researcher steering the grant that the University of California, San Francisco, was awarded to examine the utility of newborn sequencing. "We don't know if sequencing is at a point where it will be useful or not," says Puck, whose team is taking another look at dried blood spots from babies in California who had newborn screening and tested positive for metabolic diseases. Puck and the other investigators are extracting DNA from the blood spots, which are stored at −20 degrees Celsius. "We are seeing if we exome sequence these blood spots, can we find the same information or better?" says Puck.

For Puck, the professional quest turns personal at national conferences, where she connects with grieving parents. They are drawn to her like a beacon, wanting to talk about their experiences, about how helpless they felt when no one knew what was wrong with their child. Puck recalls meeting Jennifer Garcia and being impressed at how she had channeled her sorrow over Cameron's death into action, pushing Texas to include SCID on its newborn screening panel. "I remember them all," says Puck.

⌐

Sequencing all newborns to scan for disease risk throughout their lifetime sounds futuristic, but what's even more so is the idea of eliminating disease in the first place. That's what some scientists hope that CRISPR, a gene-editing technology that has sparked fear of custom-built babies, can do.

CRISPR—short for Clustered Regularly Interspaced Short Palindromic Repeats—modifies genes, enlisting a protein called

Cas9 to home in on an individual defective DNA sequence and repair the mistake. With its biologic editing prowess, CRISPR has the potential to fix many genetic diseases, a skill that has been demonstrated in mice with Duchenne muscular dystrophy and liver disease.

Researchers are also using CRISPR to try to edit cells in HIV-positive patients so that they don't go on to develop AIDS. (This approach is relevant only for a small percentage of HIV patients who will get a bone marrow transplant, but it illustrates CRISPR's potential for impact.) "Or—and this idea is decades away from execution—you could figure out which genes make humans susceptible to HIV overall," wrote the science journalist Amy Maxmen in *Wired*. "Make sure they don't serve other, more vital purposes, and then 'fix' them in an embryo. It'd grow into a person immune to the virus."

What an amazing biomedical triumph that would be—an AIDS-proof embryo. In 2016, Chinese researchers took a stab at it, using CRISPR to try to make nonviable embryos resistant to HIV. The difficulties—only some of the embryos contained the mutation that had been intended, some remained unchanged, and others acquired completely different mutations—underscored that CRISPR is a work in progress. Scientifically speaking, says Katrine Bosley, CEO of gene-editing company Editas Medicine, "it wasn't a serious attempt."

This is challenging stuff. Making certain that a gene that assists in development of disease doesn't also have some super-helpful properties that protect from other diseases, or facilitate important cellular functions, isn't so easy. We don't know the consequences of this genetic tinkering. It's like building an elaborate Lego creation, then gingerly removing one brick: you can cross your fingers that it won't undermine the integrity of your structure, but you can't know for sure until you ease out the brick.

"The technology is there, absolutely, to engineer embryos and try to tweak them and enhance them in one way or another," says Nathaniel Comfort, a professor at Johns Hopkins University who wrote *The Science of Human Perfection*. He was speaking in 2015 as part of a panel that the Center for Genetics and Society had convened to tangle with the uses and potential misuses of genetic technologies. "My guess is that it probably won't work the way people think it will. Biology always tends to throw curveballs at you. It always turns out to be a lot more complex than you thought." Consider, hypothetically, a "particular form of a gene that supposedly raises IQ 8 to 10 points. You think, who wouldn't want to have a smarter child? But we don't know what else that gene does. That gene might have all sorts of other effects we can't see."

Rather than wholesale changes, what may be more realistic in the near term is a focus on what Bosley calls "meaningful outcomes." "Of course you want to cure someone if that's possible, but for many diseases, that may not be feasible," says Bosley. "However, if you could slow or stop the progress of a disease, wouldn't that be a meaningful outcome for a patient?"

With Duchenne muscular dystrophy (DMD), which affects mostly boys, progressive muscle weakness confines kids to wheelchairs and inhibits their breathing, which leads to death by age twenty-five. We have a pretty good understanding of the genetic changes that lead to DMD, but an understanding doesn't necessarily translate into a fix. Much of the challenge has to do with how DMD affects muscle tissue; the disorder stems from a lack of dystrophin, a protein that stabilizes muscle cells.

Instead of trying to correct the flaw in every muscle fiber, perhaps a CRISPR-based medicine could be delivered to specific muscles to improve their function. "It's a challenging disease because you have a lot of muscle affected," says Bosley. "So we think about different ways that we might be helpful. If it's overly ambitious to think about getting to every muscle cell, what if we use CRISPR to do local delivery and target the diaphragm for easier breathing?"

We are sitting in Bosley's office in Kendall Square in Cambridge, Massachusetts, the heart of the biotech industry, on the first warm springlike afternoon in May. Bosley's brown eyes flash as we talk about CRISPR's potential. "It's pretty amazing what you're doing here," I say.

"It is!" she responds. "It's science fiction."

Part of what drew Bosley to Editas Medicine, which went public in a $94 million IPO in 2016, are the ethical challenges inherent in CRISPR's capabilities. Editas is working only on somatic cell disease, which is confined to cells other than egg or sperm. Changes to somatic cells aren't passed on to future generations as are changes to germ cells—egg or sperm. It's this latter prospect of amending the human germline, altering the laws of inheritance, that has unleashed a frenzy of concern.

Similar debates about germline changes echo in the discussion of mitochondrial replacement therapy (MRT), which creates an embryo using DNA from three people to avoid transmission of a small number of devastating diseases. Mitochondrial disorders are inherited from the mother; the therapy involves replacing her defective mitochondria, the powerhouses of the cell, with those of a healthy woman. The resulting embryo will have nuclear DNA—DNA contained in the nucleus, the cell's command center—from the father and the mother, plus mitochondrial DNA from a healthy donor. (Our physical and behavioral traits are encoded in our nuclear DNA; mitochondrial DNA, essential to a cell's healthy energy metabolism, also holds clues to our genetic ancestry.)

The altered DNA can be passed on to future generations, which concerns those who worry that we are conducting science experiments on babies, and delights others who are pleased that a technological workaround can produce a healthy child.

Beginning in the late 1990s, nearly twenty "three-parent" babies were born in the United States via a similar methodology, cytoplasmic transfer, before the FDA cautioned researchers that they must first seek special permission. Subsequent research has involved animals. But in late 2016, a U.S. doctor announced that he had helped a Jordanian couple conceive a child in Mexico, where there are no rules against mitochondrial replacement techniques. Dr. John Zhang of New Hope Fertility Center in New York City took the nucleus from the mother's egg and placed it into a donor egg whose nucleus had been discarded. The mother has a genetic mutation in her mitochondria that resulted in the death of her two previous infants from Leigh syndrome, a neurological disorder that is typically fatal by age three. The couple's baby was born in April 2016. Five months later, he was showing no signs of disease.

In 2015, the United Kingdom passed legislation allowing fertility specialists to perform MRT, becoming the first country to formally sanction the technique. In the United States, an expert panel assembled by the Health and Medicine Division (formerly the Institute of Medicine) of the National Academies of Science, Engineering, and Medicine recommended in 2016 that the FDA reverse course and allow researchers to conduct clinical trials, with a major caveat: at least initially, until proven safe, only male embryos should receive transferred mitochondria to sidestep any risk to the human germline. Since only women pass along mitochondrial DNA, a baby boy conceived via MRT would not convey his modified genetic material to his children. For now, the issue is moot: a federal law enacted at about the same time as the Division issued its report precludes the FDA from research "in which a human embryo is intentionally created or modified to include heritable genetic modification."

As a group, ethicists are less nervous about MRT than they are about CRISPR, because of what the expert panel calls "meaningful differences" between those genetic modifications that result from MRT and those that result from CRISPR's modification of nuclear DNA. Simply put, nuclear DNA plays a much more central role in determining a person's traits than mitochondrial DNA. (Attention, sports nuts: the Division's report did note that "while some forms of energetic 'enhancement' (such as selecting for mitochondrial DNA to increase aerobic capacity) might hypothetically be possible through MRT, they appear to be far fewer and more speculative relative to what might be possible in modifications of [nuclear] DNA.")

Speaking in one voice in 2015, scientists at an international genome-editing conclave convened to discuss CRISPR expressed their wariness over germline editing, but didn't rule out research, as long as CRISPR-manipulated embryos aren't allowed to grow into actual babies. (It's not clear who, if anyone, would be able to police and enforce such a decree.) Within weeks, a British scientist had received a green light to apply CRISPR to embryos to illuminate their early patterns of development, though the embryos would not be allowed to divide beyond seven days after conception. Without a doubt, we stand on the cusp of a world of dizzying possibilities, underscored by a statement by the prominent geneticist George Church that CRISPR, quite literally a cutting-edge technology, is already en route to becoming passé. Rather than edit genomes, Church has wondered, why not construct artificial ones? An endeavor called Human Genome Project–Write would forge human DNA from chemicals, which could in theory allow scientists to manufacture a human genome impervious to viruses and even synthesize a human being without biological parents. The prospect prompted the Stanford bioengineer Drew Endy and the Northwestern bioethicist Laurie Zoloth to wonder, "Would it be OK, for example, to sequence and then synthesize Einstein's genome? If so, how many Einstein genomes should be made and installed in cells, and who would get to make them?"

In a commentary in *Nature*, bioethicists acknowledged the bewilderment of many in their field over how exactly to impose structure and regulations on the powerful new technology embodied by CRISPR. "For decades, people have been arguing about the pros and cons of human germline modification, how to distinguish medical treatments from enhancement, what rights parents have over the lives of their children and so on. Yet good models for how to enable a diversity of perspectives to shape morally contested areas of emerging science and technology are hard to find."

It's one thing, of course, to press CRISPR into the service of modifying genes to try to cure disease, which seems more urgent, more defensible, than modifying them to build a "better" baby.

Yet regardless of the reason, the public has expressed trepidation about appropriate uses and limits for technologies like CRISPR, according to a poll of 1,000 U.S. adults on behalf of *STAT,* a health/science publication, and the Harvard School of Public Health. Eighty-three percent of those polled in 2016 said it should be illegal to edit the genes of unborn babies to boost intelligence or enhance physical characteristics. Sixty-five percent frowned on gene editing even to reduce the risk of developing serious diseases, although 44 percent responded that they would support government-funded research into embryonic gene editing focused on severe diseases such as cystic fibrosis and Huntington's disease. (If a couple is trying to avoid a genetic disease that runs in their family, it's much less technically fraught to employ the existing preimplantation genetic diagnosis technology that Deena Kobell used to create and select embryos without a specific condition than to mess around with gene editing. Unlike CRISPR, PGD doesn't change embryos; it simply identifies those that are healthy.)

The uncertainty about CRISPR's reach is so broad-based that in 2016, the U.S. director of national intelligence included gene editing on an annual worldwide threat assessment report that reflects concerns of the Central Intelligence Agency, the National Security Agency, and other covert operations. Gene editing as a potential means of creating "weapons of mass destruction and proliferation" joined a list of formidable fellow threats including ISIS terrorism and a presumed nuclear detonation by North Korea. CRISPR technology could, in theory, be used for nefarious purposes such as breeding killer mosquitoes or spreading viruses that destroy agricultural strongholds.

◻

Caution is urged on the brink of any new technology, and so it was with the birth of Connor Levy in 2013. It's not every baby whose arrival is heralded by international headlines: "IVF Baby Born Using Revolutionary Genetic-Screening Process."

Connor was lauded as the first baby in the world who began life having being screened "using a procedure that can read every letter of the human genome." The truth was more pedestrian, less sexy. Yes, he was analyzed by sequencing techniques before he ever made it to his mother's uterus. But the fuss wasn't so much about Connor, per se; it was more about the potential of the technology used to analyze his embryonic self. As you know by now, sequencing an embryo, a newborn baby, a child, or an adult could theoretically yield boundless information about health and disease risk, replete with all the accompanying moral and ethical implications. But this embryo wasn't being sequenced because Marybeth Levy, a mortgage banker, and David Levy, a nurse, were hard-charging Tiger Parents hungry for as much data as possible about their baby-to-be. Rather, he was sequenced as part of a research study that used the technology instead of existing genetic techniques to screen for a chromosomally normal embryo. Recall that embryos with abnormal chromosomes are less likely to result in the birth of a baby.

Main Line Fertility in Bryn Mawr, Pennsylvania, where the Levys were seeking treatment, biopsied the couple's thirteen embryos, siphoning off a few cells from each and shipping them to researcher Dagan Wells at the University of Oxford. Sequencing revealed that only three embryos had the correct number of chromosomes. One was transferred to Marybeth and became Connor, who was munching grilled cheese while playing peek-a-boo behind my laptop when I visited the Levys at their narrow row house in a gritty part of Philadelphia. Connor was seventeen months old at the time, with spiky blond hair and big blue eyes. One could be forgiven for thinking that sequencing must have identified an adorable-baby gene in Connor. He is that cute.

His parents were stunned by the buzz their story had generated. They appeared on the talk show *The Doctors* and were contacted by reporters from Spain, Israel, and Germany. At work, colleagues regularly ask Marybeth to see pictures of the "famous baby." "If you Google his name, you see all the stuff that comes up," says Marybeth. "People say it's unethical and the beginning of a super-human race. They say we're going to eliminate brown-haired, brown-eyed babies."

The Levys have nothing against brown eyes or brown hair. All they wanted was a child after failing to conceive on their own. "He eats, he sleeps, he's healthy," says David. "We joke about how great it is to have a genetically enhanced baby."

Technology, of course, continues to evolve, to grow more so-phisticated, to yield more and better information about our children before they're conceived, while they're cushioned in a watery womb and once they emerge into this world of boundless possibility.

In truth, although technologies now exist that our parents couldn't have envisioned, the goal of most mothers and fathers remains the same: a healthy baby. Technology is just a means to an end, a way to make—and keep—children healthy. As parents, we are first and foremost protectors, of our children's health and of their futures—the mythical "perfect child" be damned.

Marybeth's fertility doctor, Michael Glassner, told the Levys that they had "hit the genetic lottery" when the stars aligned to tap an embryonic Connor for sequencing. But the Levys see the little boy they laughingly refer to as "some type of Frankenstein" differently. Like parents the world over, regardless of whether they are taking advantage of evolving genetic technologies, they are "just happy to have a healthy kid."

Notes

Introduction

Much of the material in this chapter is derived from interviews with Ronald Wapner and his genetic counselors and patients, September 2013.

3 *with about sixty new mutations*: Donald F. Conrad et al., "Variation in Genome-Wide Mutation Rates Within and Between Human Families," *Nature Genetics* 43, no. 7 (2011): 712–14.

4 *inversion 9, a transposed ninth chromosome*: Shinichiro Nanko, "Schizophrenia with Pericentric Inversion of Chromosome 9: A Case Report," *Japanese Journal of Psychiatry and Neurology* 47, no. 1 (1993): 47–49.

5 *tests that transcended the boundaries*: Kenneth P. Tercyak et al., "Parents' Attitudes Toward Pediatric Genetic Testing for Common Disease Risk," *Pediatrics* 127, no. 5 (2011): 1288–95.

6 *As McBride characterized the parents*: Bonnie Rochman, "Genetic Testing for Kids: Is It a Good Idea?," Time.com, April 18, 2011.

7 *survey revealed overwhelming support*: Susan E. Waisbren et al., "Parents Are Interested in Newborn Genomic Testing During the Early Postpartum Period," *Genetics in Medicine* 17, no. 6 (2015): 501–504.

7 *nonchalant teenagers want a piece of the action*: Sara G. Miller, "Teens Want to Know Genetic Test Results," *Live Science*, October 9, 2015.

8 *one geneticist I interviewed*: Geneticist who requested anonymity, interview with the author, May 2015.

8 *"If the barley grows"*: "A Timeline of Pregnancy Testing," "A Thin Blue Line: The History of the Pregnancy Test Kit," Office of NIH History, https://history.nih.gov/exhibits/thinblueline/timeline.html.

9 *website cleverly dubs Baby's First Test*: "Screening Procedures," Baby's First Test, http://www.babysfirsttest.org/newborn-screening/screening-procedures.

10 *a cheap, effective way to screen for it*: "Promoting Safe and Effective Genetic Testing in the United States," National Human Genome Research Institute, https://www.genome.gov/10002397/.

10 *newborn screening has saved countless lives*: Over the years, newborn screening has expanded to include every state and at least thirty-four diseases, which make up the core conditions that comprise the federal Recommended Uniform Screening Panel (RUSP). Disorders on the panel are severe, cost-effective to screen for, and treatable. See http://www.babysfirsttest.org/newborn-screening/the-recommended-uniform-screening-panel, http://www.hrsa.gov/advisorycommittees/mchbadvisory/heritabledisorders/recommendedpanel/index.html.

Adding a disorder to the RUSP is time-consuming and complex. To start the process, someone—a parent whose child died, perhaps, or a researcher or advocacy group—must nominate a condition for inclusion. The Advisory Committee on Heritable Disorders in Newborns and Children, a group of doctors and specialists in genetics and pediatrics, evaluates each nomination. The committee's quarterly meetings are open to the public, and it's not unusual for grieving parents to attend to share personal stories of infants lost to diseases for which screening was not offered. See http://www.hrsa.gov/advisorycommittees/mchbadvisory/heritabledisorders/nominatecondition/index.html.

States are not required to follow the recommendations. Newborn screening is considered a state responsibility, so each state's public-health department is free to decide for itself which conditions it will include on its disease panel. As a result, different states screen for different conditions.

In other words, despite the presence of lots of smart people on the committee who have devoted their lives to researching the best ways to detect newborn illness, the recommendations are not binding. In an ideal world, states would screen at minimum for the list of recommended conditions. Many do. But not all states budget for that. Especially when new conditions are added, individual states can take a while to fall into line.

10 *screening identifies more than 5,000 infants*: "About Newborn Screening: Screening Facts," Baby's First Test, http://www.babysfirsttest.org/newborn-screening/screening-facts.

12 *In a pivotal study published in 2012*: Ronald J. Wapner et al., "Chromosomal Microarray Versus Karyotyping for Prenatal Diagnosis," *New England Journal of Medicine* 367 (2012): 2175–84.

12 *Some of these changes are meaningless*: Wapner found that even when women had normal ultrasounds and karyotype results—generally a reason to exhale a sigh of relief and start scouting paint swatches for the nursery—they might have cause to worry: in his study, 1.7 percent of these women actually were

pregnant with babies with undetected DNA deletions or duplications—what Wapner calls a "clinically relevant" finding. And about 1 percent of women will have a "finding of uncertain significance," in which doctors have no idea what the identified deletions or duplications signify. They're the experts, but they're in the dark.

13 *lucky enough to be Angelina Jolie*: Angelina Jolie, "My Medical Choice," *The New York Times*, May 14, 2013.

14 *Jolie has three biological children*: Bonnie Rochman, "Hard Choices Angelina Jolie Faces About Testing Her Kids for Breast Cancer Genes," Time.com, May 14, 2013.

1. How the Jews Beat Tay-Sachs

Much of the material in this chapter is derived from interviews with Shivani Nazareth via phone and on-site at Counsyl in January 2016; with Brittany Madore, with Peter Kasdan, with Sophie-Shifra Gold, and with Mike Kaback. Joanna Zeiger and her parents deferred to Kaback to recount the story of Joanna's role in developing Tay-Sachs screening.

17 *children conceived by a couple who are both carriers*: "Why Is a Baby Born with Tay-Sachs Disease?," National Tay-Sachs and Allied Diseases, http://www .tay-sachs.org/taysachs_disease.php.

18 *One of every 27 Ashkenazi Jews*: "Who Is at Risk?," National Tay-Sachs and Allied Diseases, http://www.tay-sachs.org/taysachs_disease.php.

19 *markets its comprehensive panel as a gift*: "Genetic Testing? Now That's a Gift!," JScreen in association with Emory University, https://jscreen.org /gift/.

19 *One company, Gene by Gene Ltd.*: "A Carrier Screen to Help You Grow Your Family with Confidence," Gene by Gene, https://gxg.genebygene.com /carrier-screening/.

19 *a test that screens for more than 250 diseases*: Bonnie Rochman, "New Genetic Tests for Women Who Are Expecting," *The Wall Street Journal*, January 5, 2015.

20 *DNA is our vehicle of heredity*: If a picture is indeed worth a thousand words, the discovery of DNA's classic twisty ladder shape is encyclopedic in impact. In 1953, the American biologist James Watson and the English physicist Francis Crick announced that they knew what DNA looked like and proceeded to sketch out its three-dimensional double helix structure (without so much as a shout-out to Rosalind Franklin, the female scientist who did much of the work that led to their conclusion).

21 *bodies grow by making new cells*: "How Does DNA Change?," My46, University of Washington, https://www.my46.org/intro/how-does-dna-change.

21–22 *One of every 280 babies born*: Shivani B. Nazareth et al., "Changing Trends in Carrier Screening for Genetic Disease in the United States," *Prenatal Diagnosis* 35, no. 10 (2015): 931–35.

23 *Madore described her experience*: "Sullivan's Life Sheds Light on SMA," Counsyl Blog, August 15, 2013, https://blog.counsyl.com/2013/08/15/sullivans-life-sheds-light-on-sma/.

24 *one in fifty Americans is an SMA carrier*: "How Common Is Spinal Muscular Atrophy?," Counsyl, https://www.counsyl.com/services/family-prep-screen/diseases/spinal-muscular-atrophy/.

25 *If such an error isn't fixed*: "What Is Mutation?," Genetic Science Learning Center, University of Utah, http://learn.genetics.utah.edu/content/variation/mutation/.

25 *Other mutations may emerge*: "What Is a Gene Mutation and How Do Gene Mutations Occur?," Genetics Home Reference, U.S. National Library of Medicine, https://ghr.nlm.nih.gov/primer/mutationsanddisorders/genemutation.

25 *They discovered that the average person*: Yali Xue et al., "Deleterious- and Disease-Allele Prevalence in Healthy Individuals: Insights from Current Predictions, Mutation Databases, and Population-Scale Resequencing," *The American Journal of Human Genetics* 91, no. 6 (2012): 1022–32.

26 *There are all sorts of triggers*: "What Is Personal Genetics?," Personal Genetics Education Project, Department of Genetics, Harvard Medical School, http://www.pged.org/personal-genetics-101/what-is-personal-genetics/.

26 *as with a disease such as adult-onset diabetes*: Laura Dean and Jo McEntyre, *The Genetic Landscape of Diabetes* (Bethesda, MD: National Center for Biotechnology Information, 2004), chapter 3, "Genetic Factors in Type 2 Diabetes," http://www.ncbi.nlm.nih.gov/books/NBK1665/.

28 *Karen was well into her second pregnancy*: Paul McMullen, "Before Start, Mind Races," *The Baltimore Sun*, September 9, 2000.

28 *they decided they would not see their infant*: Robert J. Desnick and Michael M. Kaback, *Tay-Sachs Disease*, Advances in Genetics, vol. 44 (Academic Press, 2001), 253–66.

30 *Tay-Sachs is caused by a deficiency*: "Learning About Tay-Sachs Disease," National Human Genome Research Institute, https://www.genome.gov/10001220/learning-about-taysachs-disease/.

30–31 *place fourth in the triathlon competition*: Joanna Zeiger's curriculum vitae, http://joanna-zeiger.com/jzeiger/academic-cv/.

33 *more than 50,000 carriers were identified*: Michael Kaback et al., "Genetic Screening in the Persian Jewish Community: A Pilot Study," *Genetics in Medicine* 12, no. 10 (2010): 628–33.

33 *screening programs have reduced the incidence*: Ibid.

34–35 *the carrier rate in this community*: Gaucher disease, National Gaucher Foundation, Inc., http://www.gaucherdisease.org.

35 *The American College of Obstetricians and Gynecologists (ACOG) recommends*: "Preconception and Prenatal Carrier Screening for Genetic Diseases in Individuals of Eastern European Jewish Descent," Committee on Genetics, ACOG Committee Opinion, no. 442, October 2009, reaffirmed 2014, http://www.acog.org/Resources-And-Publications/Committee-Opinions /Committee-on-Genetics/Preconception-and-Prenatal-Carrier-Screening -for-Genetic-Diseases-in-Individuals-of-Eastern-European; and Susan J. Gross et al., "Carrier Screening in Individuals of Ashkenazi Jewish Descent," *Genetics in Medicine* 10, no. 1 (2008): 54–56.

35 *screen for about nineteen disorders*: Bonnie Rochman, "New Genetic Tests for Women Who Are Expecting," *The Wall Street Journal*, January 5, 2015.

36 *a resolution was passed*: Resolution adopted by the Central Conference of American Rabbis at their 86th annual convention, 1975, http://ccarnet.org /rabbis-speak/resolutions/1975/genetic-diseases-jewish-1975/.

37 *more likely to occur in the Sephardic population*: "All About Genetic Diseases That Strike Sephardic Jews," *The Forward*, August 9, 2014

38 *a "compatibility check"*: Dor Yeshorim, "Process: How It Works," http·// doryeshorim.org/process/.

38 *Jews from Iraq, for example*: "Recommendations for Genetic Testing," State of Israel Ministry of Health, http://www.health.gov.il/English/Topics/Genetics /checks/Pages/GeneticTestingRecommendations.aspx.

38 *Israel covers prenatal genetic testing*: "Genetic Investigation and Tests Before or During Pregnancy," State of Israel Ministry of Health, http://www.health .gov.il/English/Topics/Genetics/checks/screening-genes/Pages/default .aspx.

39 *half, in fact—get pregnant accidentally*: Fact Sheet: "Unintended Pregnancy in the United States," Guttmacher Institute, March 2016, https://www .guttmacher.org/fact-sheet/unintended-pregnancy-united-states.

42 *carriers for just one disease: cystic fibrosis*: "Update on Carrier Screening for Cystic Fibrosis," Committee on Genetics, ACOG Committee Opinion, no. 486, April 2011, reaffirmed 2014, http://www.acog.org/Resources -And-Publications/Committee-Opinions/Committee-on-Genetics /Update-on-Carrier-Screening-for-Cystic-Fibrosis.

42 *ACOG and other medical societies released*: Janice G. Edwards et al., "Expanded Carrier Screening in Reproductive Medicine—Points to Consider: A Joint Statement of the American College of Medical Genetics and Genomics, American College of Obstetricians and Gynecologists, National Society of Genetic Counselors, Perinatal Quality Foundation, and Society for Maternal-Fetal Medicine," *Obstetrics and Gynecology* 125, no. 3 (2015): 653–62.

42 *diseases that make up the ever-expanding testing panels*: Ingrid Lobo, "Same Genetic Mutation, Different Genetic Disease Phenotype," *Nature Education* 1, no. 1 (2008): 64.

43 *DNA is not necessarily destiny*: A fascinating study, the Resilience Project, aims to track down what its researchers are calling "unexpected heroes"—people who have variants strongly identified with genetic diseases, many of them devastating, that typically manifest in childhood. Yet these heroes have reached the age of forty and beyond without ever developing the condition. In many cases, these disease defeaters may have no clue that they even harbor the particular variant. The goal of the study is to identify these fortunate folks and analyze their DNA for other changes that mitigate or "buffer" the disease-associated variant.

43 *a 65 percent chance of developing the disease*: "BRCA1 and BRCA2: Cancer Risk and Genetic Testing," National Cancer Institute, http://www.cancer.gov/about-cancer/causes-prevention/genetics/brca-fact-sheet.

2. Playing God

Much of the material in this chapter is derived from interviews with Susan Davis, Jennifer Davis, Deena Kobell, Mark Hughes, Gwen Quinn, Sue Friedman, and Paul Lombardo.

45 *increasing their risk of getting breast cancer*: "BRCA1 and BRCA2: Cancer Risk and Genetic Testing," National Cancer Institute, http://www.cancer.gov/about-cancer/causes-prevention/genetics/brca-fact-sheet.

45 *had a crude mastectomy*: The radical mastectomy was introduced in 1882 by William Halsted, a professor of surgery at Johns Hopkins University. Though Halsted was the first to perform a radical mastectomy, which involves removal of all breast tissue, axillary lymph nodes, and pectoralis muscles, the origins of this surgery stem from *Chirurgie*, a surgical compendium written by the German physician Lorenz Heister. For close to a century, it was the standard of care no matter the patient's age or type of tumor. Treatment today focuses on breast-conserving surgery. See http://www.amsj.org/archives/3019, http://www.cancer.org/cancer/cancerbasics/thehistoryofcancer/the-history-of-cancer-cancer-treatment-surgery.

46 *Mary-Claire King would lead the way*: "BRCA1," Genetics Home Reference, U.S. National Library of Medicine, http://ghr.nlm.nih.gov/gene/BRCA1.

46 *they had a name for it*: "Mary-Claire King, PhD: ASHG President (2012)," The American Society for Human Genetics, http://www.ashg.org/press/mcking_bio.shtml.

46 *BRCA2 was discovered*: "BRCA2," Genetics Home Reference, U.S. National Library of Medicine, http://ghr.nlm.nih.gov/gene/BRCA2.

46 *More than 1,600 individual mutations*: Nancie Petrucelli et al., "Hereditary Breast and Ovarian Cancer Due to Mutations in BRCA1 and BRCA2," *Genetics in Medicine* 12, no. 5 (2010): 245–59.

48 *more BRCA carriers are considering*: Bonnie Rochman, "Family with a Risk of Cancer Tries to Change Its Destiny," *The Wall Street Journal*, February 17, 2014.

49 *doesn't routinely recommend that PGD be discussed*: "Genetic/Familial High-Risk Assessment: Breast and Ovarian," National Comprehensive Cancer Network Clinical Practice Guidelines in Oncology, Version 2.2016.

49 *One study that examined knowledge*: Gwendolyn Quinn et al., "Attitudes of High-Risk Women Toward Preimplantation Genetic Diagnosis," *Fertility and Sterility* 91, no. 6 (2009): 2361–68.

50 *she'd have a one-in-two chance*: "How Do People Get Breast or Ovarian Cancer?," Breast Cancer and Ovarian Cancer, Genetic Science Learning Center, University of Utah, http://learn.genetics.utah.edu/content/disorders /multifactorial/brca/.

51 *the first-ever use of PGD*: Alan H. Handyside et al., "Birth of a Normal Girl After In Vitro Fertilization and Preimplantation Diagnostic Testing for Cystic Fibrosis," *New England Journal of Medicine* 327, no. 13 (1992): 905–909.

52 *Eve's risk is on par*: "Breast Cancer Risk in American Women," National Cancer Institute, http://www.cancer.gov/types/breast/risk-fact-sheet.

53 *In response to an article I wrote*: Bonnie Rochman, "Family with a Risk of Cancer Tries to Change Its Destiny," *The Wall Street Journal*, February 17, 2014.

53 *men with a breast cancer mutation*: Couples in which one partner has a mutation may decide to select for gender. They want a son because even if that boy has a *BRCA* mutation, his risk of getting cancer as an adult, while higher than that of the average male, is still relatively low compared to a woman with a mutation.

53 *As the bioethicist Arthur Caplan has noted*: "Flunking the Genetic Test: Why We're Not Ready for Mass Genetic Screening and Testing," *CAP (College of American Pathologists) Today*, January 2003, http://www.captodayonline.com /Archives/feature_stories/geneticTesting.html.

54 *Quinn conducted a large survey*: Susan T. Vadaparampil, "Factors Associated with Preimplantation Genetic Diagnosis Acceptance Among Women Concerned About Hereditary Breast and Ovarian Cancer," *Genetics in Medicine* 11, no. 10 (2009): 757–65.

54 *a focus group of thirteen women*: Gwendolyn P. Quinn et al., "*BRCA* Carriers' Thoughts on Risk Management in Relation to Preimplantation Genetic Diagnosis and Childbearing: When Too Many Choices Are Just as Difficult as None," *Fertility and Sterility* 94, no. 6 (2010): 2473–75.

55 *PGD has its roots*: Alan Handyside, "Celebrating 20 Years of Preimplantation Genetic Diagnosis," *BioNews*, July 23, 2010, http://www.bionews.org.uk /page_67228.asp.

56 *to avoid passing on an X-linked disorder*: "X-Linked Recessive Genetic Defects— How Girls Are Affected," MedlinePlus, U.S. National Library of Medicine, https://www.nlm.nih.gov/medlineplus/ency/imagepages/19096.htm.

56 *they performed PGD for cystic fibrosis*: Alan H. Handyside et al., "Birth of a Normal

Girl After In Vitro Fertilization and Preimplantation Diagnostic Testing for Cystic Fibrosis," *New England Journal of Medicine* 327, no. 13 (1992): 905–909.

57 *better treatments have increased longevity*: "Cystic Fibrosis: Life Expectancy," National Jewish Health, http://www.nationaljewish.org/healthinfo/conditions /cysticfibrosis/life-expectancy/.

57 *a hub for "reproductive tourism"*: Sumathi Reddy, "Fertility Clinics Let You Select Your Baby's Sex," *The Wall Street Journal*, August 17, 2015.

57 *an online survey of U.S. fertility clinics*: Susannah Baruch et al., "Genetic Testing of Embryos: Practices and Perspectives of U.S. In Vitro Fertilization Clinics," *Fertility and Sterility* 89, no. 5 (2008): 1053–58.

57 *"no ethical obligation to provide"*: "Use of Reproductive Technology for Sex Selection for Nonmedical Reasons," Ethics Committee opinion, American Society for Reproductive Medicine, *Fertility and Sterility* 103, no. 6 (2015): 1418–22.

57 *opposes the use of PGD to select embryos*: "Sex Selection," Committee on Ethics, ACOG Committee Opinion, no. 360, *Obstetrics and Gynecology* 109, no. 2 part 1 (2007): 475–78.

58 *the connection between genes and intelligence*: Jon D. Arbogast, "Progress on All Fronts in Race to Map Genes; Why Do Research?," *The New York Times*, May 17, 1998.

59 *"Nobel Prize sperm bank"*: David Plotz, "The Myths of the Nobel Sperm Bank," *Slate*, February 23, 2001.

60 *"traditional reproduction may begin to seem antiquated"*: Gregory Stock, *Redesigning Humans: Choosing Our Genes, Changing Our Future* (New York: Houghton Mifflin, 2002), 55.

61 *collects and shares genome data*: "What Is the Harvard PGP?," Harvard Personal Genome Project, https://github.com/PGPHarvard/FAQ/wiki/FAQ.

61 *more than forty repeats*: Nicholas T. Potter et al., "Technical Standards and Guidelines for Huntington Disease," American College of Medical Genetics, 2006, https://www.acmg.net/pages/acmg_activities/stds-2002/hd.htm.

62 *a deaf lesbian couple*: Merle Spriggs, "Lesbian Couple Create a Child Who Is Deaf Like Them," *Journal of Medical Ethics* 28, no. 5 (2002): 283.

62 *3 percent of PGD clinics surveyed*: Susannah Baruch et al., "Genetic Testing of Embryos: Practices and Perspectives of U.S. In Vitro Fertilization Clinics," *Fertility and Sterility* 89, no. 5 (2008): 1053–58.

62 *doctors don't sanction PGD for that purpose*: Darshak M. Sanghavi, "Wanting Babies Like Themselves, Some Parents Choose Genetic Defects," *The New York Times*, December 5, 2006.

62 *Dwarfism rates vary*: "Achondroplasia," Genetics Home Reference, U.S. National Library of Medicine, https://ghr.nlm.nih.gov/condition/achondroplasia #statistics.

63 *fatal "double-dominant" mutation*: Darshak M. Sanghavi, "Wanting Babies Like Themselves, Some Parents Choose Genetic Defects," *The New York Times*, December 5, 2006.

63 *first documented case of PGD for dwarfism*: Ibid.
64 *legally enshrined accommodations*: Dan Kennedy, *Little People: Learning to See the World Through My Daughter's Eyes* (Emmaus, PA: Rodale Books, 2003), chapter 12.
65 *"The best men must have intercourse"*: Calum MacKellar and Christopher Bechtel, eds., *The Ethics of the New Eugenics* (New York and Oxford: Berghahn Books, 2014), 15.
65 *evidence in favor of culling unsuitables*: Paul A. Lombardo, "Return of the Jukes: Eugenic Mythologies and Internet Evangelism," *Journal of Legal Medicine* 33, no. 2 (2012): 207–33.
65 *things started to get competitive*: Steven Selden, "Transforming Better Babies into Fitter Families: Archival Resources and the History of the American Eugenics Movement, 1908–1930," *Proceedings of the American Philosophical Society* 149, no. 2 (2005), 199–225.
67 *"represented in court by the evidence"*: Paul A. Lombardo, "Disability, Eugenics, and the Culture Wars," *Saint Louis University Journal of Health Law and Policy* 2 (2008): 57–79.
68 *Even Kennedy admits*: Dan Kennedy, *Little People: Learning to See the World Through My Daughter's Eyes* (Emmaus, PA: Rodale Books, 2003), 248–49.
69 *just part of what makes a person tick*: Steven Selden, "Transforming Better Babies into Fitter Families: Archival Resources and the History of the American Eugenics Movement, 1908–1930," *Proceedings of the American Philosophical Society* 149, no. 2 (2005), 199–225.

3. The Other Scarlet "A"
Much of the material in this chapter is derived from interviews with patients at Ron Wapner's clinic and with Brian Skotko, Megan Allyse, Mary-Frances Garber, Jen Sipress, Addie Morfoot, Elizabeth Nash, Warren Hern, Ricki Lewis, and William Schoolcraft.

74 *microarray would reveal all sorts of genetic blips*: "What Does CMA Detect?," Genomic Tools: Chromosomal Microarray, National Coalition for Health Professional Education in Genetics, http://www.nchpeg.org/microarray/what-does-cma-detect.
74 *1 of 100 pregnancies results*: "Prenatal Testing for Down Syndrome," UCSF Medical Center, https://www.ucsfhealth.org/education/down_syndrome/.
75 *the legalization of abortion*: Ilana Löwy, "Prenatal Diagnosis: The Irresistible Rise of the 'Visible Fetus,'" *Studies in History and Philosophy of Science Part C: Studies in History and Philosophy of Biological and Biomedical Sciences* 47, part B (2014): 290–99.
76 *One of every 792 babies*: Gert De Graaf et al., "Estimates of the Live Births, Natural Losses, and Elective Terminations with Down Syndrome in the

United States," *American Journal of Medical Genetics Part A* 167, no. 4 (2015): 756–67.

76 *the smallest chromosome*: "Chromosome 21," Genetics Home Reference, U.S. National Library of Medicine, https://ghr.nlm.nih.gov/chromosome/21.

76 *expanded its prenatal screening recommendations*: "Prenatal Diagnostic Testing for Genetic Disorders," Practice Bulletin 162, American College of Obstetricians and Gynecologists' Committee on Practice Bulletins—Obstetrics, Committee on Genetics, and the Society for Maternal-Fetal Medicine, May 2016.

76 *comprehensive look at Down syndrome live-birth rates*: Gert De Graaf et al., "Estimates of the Live Births, Natural Losses, and Elective Terminations with Down Syndrome in the United States," *American Journal of Medical Genetics Part A* 167, no. 4 (2015): 756–67.

77 *its greater accuracy combined with its ease of use*: "Prenatal Cell-Free DNA Screening: Q&A for Healthcare Providers," National Society of Genetic Counselors, http://nsgc.org/page/non-invasive-prenatal-testing-healthcare-providers.

77 *the concept of accuracy itself is nuanced and complex*: Accuracy can be complicated. In general conversation, "accuracy" means being error-free. Scientifically speaking, it actually comprises two statistical ideas that measure the likelihood of false positives and false negatives: sensitivity and specificity. The sensitivity rate refers to how many women who are carrying a fetus with Down syndrome will actually be identified. The genetic counselor Katie Stoll offers up the example of 100,000 thirty-five-year-old women who have an incidence rate of 1 in 250, which means 400 of them will be pregnant with a fetus with Down syndrome. According to NIPS labs' sensitivity rate of 99.5 percent, 398 of these women will receive a positive result, while two will receive a negative result, which will in fact be wrong. The labs also boast up to a 99.9 percent specificity rate, which measures the percentage of women who aren't carrying a Down syndrome fetus who will receive a negative test result. Applying this percentage to Stoll's hypothetical 99,600 women who don't have Down syndrome pregnancies, 99,500 of them would receive negative reports. That leaves 100 women who would get positive results when in fact their pregnancies are just fine. The likelihood of having a false-positive result rises as the incidence rate plummets.

Positive predictive value, or PPV, also plays a significant role in interpreting results. PPV considers maternal age, gestational age, and screening findings. Simply put, PPV indicates the chance that a positive screen result truly reflects a diagnosis. PPV is affected by the prevalence of disease within a population, and to this extent, it includes maternal age in the calculations. Older women can have a higher PPV because they have a greater chance of having a child with Down syndrome. A PPV for a forty-two-year-old woman who opts for NIPS will be very different from a PPV for a twenty-two-year-old woman, although the sensitivity/specificity of NIPS overall remains high.

Sensitivity, specificity, and PPV each factor into determining accuracy. See https://thednaexchange.com/2013/07/11/guest-post-nips-is-not-diagnostic -convincing-our-patients-and-convincing-ourselves/, http://www.downsyn dromeprenataltesting.com/how-accurate-is-the-new-blood-test-for-down -syndrome/.

77 *has spawned a $500 million industry*: "Non-Invasive Prenatal Testing (NIPT) Market . . . : Global Industry Analysis, Size, Volume, Share, Growth, Trends and Forecast 2014–2022," Transparency Market Research, 2015, http:// www.transparencymarketresearch.com/noninvasive-prenatal-diagnostics -market.html.

78 *new screening hasn't altered outcomes much*: Steven L. Warsof et al., "Overview of the Impact of Noninvasive Prenatal Testing on Diagnostic Procedures," *Prenatal Diagnosis* 35, no. 10 (2015): 972–79.

78 *responsible for the birth of the first IVF baby*: "Infertility Advances at the Jones Institute," Eastern Virginia Medical School, http://www.evms.edu/patient _care/specialties/jones_institute_for_reproductive_medicine/about_us/.

78 *NIPS is not perfect*: "Cell-Free DNA Screening for Fetal Aneuploidy," ACOG Committee Opinion, no. 640, September 2015, American College of Obstetricians and Gynecologists Committee on Genetics and Society for Maternal-Fetal Medicine, http://www.acog.org/Resources-And-Publications /Committee-Opinions/Committee-on-Genetics/Cell-free-DNA -Screening-for-Fetal-Aneuploidy.

78 *an underlying maternal cancer diagnosis*: Diana W. Bianchi, "Pregnancy: Prepare for Unexpected Prenatal Test Results," *Nature* 522 (2015): 29–30.

78 *terminating pregnancies they believed were affected*: Beth Daley, "Oversold Prenatal Tests Spur Some to Choose Abortions," *The Boston Globe*, December 14, 2014.

78 *positive results need to be confirmed*: "Cell-Free DNA Screening for Fetal Aneuploidy," ACOG Committee Opinion, no. 640, September 2015, American College of Obstetricians and Gynecologists' Committee on Genetics and the Society for Maternal-Fetal Medicine, http://www.acog.org/Resources-And -Publications/Committee-Opinions/Committee-on-Genetics/Cell-free -DNA-Screening-for-Fetal-Aneuploidy.

78 *there has been little public conversation*: Some companies have also expanded the screens to include genetic conditions about which families and their doctors may be unaware. Families may think they're screening for Down syndrome when in actuality they will also receive results for various "microdeletions," small bits of missing chromosomes, or sex chromosome abnormalities, a host of disorders that involve errors in the heredity of the twenty-third pair of chromosomes. When Ron Wapner noticed this trend in 2014, he was taken aback. "They didn't even tell us," says Wapner. "We said, 'You can't not inform the

patient.'" Now, many companies have made microdeletion results "opt in" or "opt out." Still, says Shivani Nazareth of Counsyl, "what's not happening is a frank discussion with patients about what else they are being screened for other than Down syndrome, and there aren't enough genetic counselors to truly provide informed consent."

78 *"consequences of the transformation of every fetus"*: Ilana Löwy, "Prenatal Diagnosis: The Irresistible Rise of the 'Visible Fetus,'" *Studies in History and Philosophy of Science Part C: Studies in History and Philosophy of Biological and Biomedical Sciences* 47 (2014): 290–99.

79 *In an op-ed in* The New York Times: Patricia Volk, "The T.M.I. Pregnancy," *The New York Times*, June 4, 2014.

79 *two letters to the editor*: "Prenatal Testing, Along with the Worries," *The New York Times*, June 14, 2014.

80 *"Collective fiction" is a term*: Nancy Anne Press, Carole H. Browner, "'Collective Fictions': Similarities in Reasons for Accepting Maternal Serum Alpha-Fetoprotein Screening Among Women of Diverse Ethnic and Social Class Backgrounds," *Fetal Diagnosis and Therapy* 8, suppl. 1 (1993): 97–106.

82 *genetic counselors who are salaried employees*: Andrew Pollack, "Conflict Potential Seen in Genetic Counselors," *The New York Times*, July 13, 2012.

84 *a phenomenon called "variable expressivity"*: "What Are Reduced Penetrance and Variable Expressivity?," Genetics Home Reference, U.S. National Library of Medicine, https://ghr.nlm.nih.gov/primer/inheritance/penetranceexpres sivity.

85 *"I'd rather my child not have this disability"*: Bonnie Rochman, "Why Down Syndrome Is on the Decline," Time.com, February 17, 2012.

86 *"If public health espouses goals"*: Adrienne Asch, "Prenatal Diagnosis and Selective Abortion: A Challenge to Practice and Policy," *American Journal of Public Health* 89, no. 11 (1999): 1649–57.

86 *"To answer Adrienne Asch's critique"*: Dan Kennedy, *Little People: Learning to See the World Through My Daughter's Eyes* (Emmaus, PA: Rodale Books, 2003), 237.

87 *couple was awarded nearly $3 million*: Aimee Green, "Jury Awards Nearly $3 Million to Portland-Area Couple in 'Wrongful Birth' Lawsuit Against Legacy Health," *The Oregonian*, March 9, 2012.

87 *"Wrongful birth" lawsuits are allowed*: Kathy Lohr, "Should Parents Be Able to Sue for 'Wrongful Birth'?" National Public Radio, May 15, 2012.

87 *One blogger wrote of the Levys*: Cassy Fiano, "Couple Wins Down Syndrome 'Wrongful Birth' Suit," *Live Action News*, March 11, 2012.

88 *Median survival is 37.5 years*: "Cystic Fibrosis: Life Expectancy," National Jewish Health, http://www.nationaljewish.org/healthinfo/conditions/cysticfibrosis/life -expectancy/ and "Cystic Fibrosis," Genetic Science Learning Center, University of Utah, http://learn.genetics.utah.edu/content/disorders/singlegene/cf/.

89 *When Morfoot shared her experience*: Addie Morfoot, "Our Impossible Parenting Choice," *Salon*, October 8, 2013.

89 *first state to pass a law*: "Chapter 14-02.1 Abortion Control Act," North Dakota Legislative Branch, http://www.legis.nd.gov/cencode/t14c02-1.pdf#nameddest =14-02p1-01; and Bonnie Rochman, "Pro-Choice or No Choice? North Dakota Wants to Ban Abortion for Fetal Abnormalities," Time.com, March 25, 2013.

90 *would have prohibited abortion on the basis of race, gender, or disability*: House Enrolled Act 1337, Section 22, IC 16-34-4, Indiana General Assembly, 2016 Session, https://iga.in.gov/static-documents/5/1/b/5/51b52d50/HB1337.05 .ENRS.pdf.

91 *women face conflicting pressures*: After *The New York Times* reported in 2015 on an Ohio bill to prohibit abortion after a prenatal diagnosis, Julia in New York City shared the decision-making process of her mother, who had a sibling with Down syndrome and decided to abort a fetus with the condition—a fetus that would have been Julia's sibling.

   "I have no regrets at all and never have," she quoted her mother as saying. "Sadness, yes, when I opened the post-op report and read about a 5 cm left foot and that it had been a boy, but not even a twinge that I chose to end this potential life in order to make my life, and those of my already born daughter and son who arrived two years later, a family that was not broken, tortured, and hell to live in as my own had been." See http://takingnote.blogs.nytimes .com/2015/09/09/down-syndrome-and-abortion-readers-share-their -stories/?smid=fb-nytimes&smtyp=cur.

91 *At least fourteen states prohibit abortion*: "State Policies on Later Abortions," Guttmacher Institute, July 1, 2016, https://www.guttmacher.org/state-policy /explore/state-policies-later-abortions.

91 *1 percent of abortions take place at 21 weeks*: "Induced Abortion in the United States," Guttmacher Institute, May 2016, https://www.guttmacher.org/fact -sheet/induced-abortion-united-states.

92 *160 different fetal disorders*: Warren M. Hern, "Fetal Diagnostic Indications for Second and Third Trimester Outpatient Pregnancy Termination," *Prenatal Diagnosis* 34, no. 5 (2014): 438–44.

93 *women receive written information*: Mark W. Leach, "The Down Syndrome Infor- mation Act: Balancing the Advances of Prenatal Testing Through Public Policy," *Intellectual and Developmental Disabilities* 54, no. 2 (2016): 84–93.

93 *the stipulation is known as "Chloe's Law"*: Marie McCullough, "Pa. Law Man- dates Information with Down Syndrome Diagnosis," *The Philadelphia Inquirer*, August 20, 2014.

94 *telling doctors they must share certain brochures*: Tara Murtha, "Pennsylvania Law Requires Doctors to Read Scripts to Pregnant Patients with Prenatal Down Syndrome Diagnoses (Updated)," *Rewire*, July 25, 2014.

94 *A reporter writing about the pamphlet*: Marie McCullough, "Pa. Law Mandates Information with Down Syndrome Diagnosis," *The Philadelphia Inquirer*, August 20, 2014.

94 *the bioethicist Arthur Caplan has argued*: Arthur L. Caplan, "Chloe's Law: A Powerful Legislative Movement Challenging a Core Ethical Norm of Genetic Testing," *PLOS Biology* 13, no. 8 (2015): e1002219.

95 *"more positively spin life with trisomy 21 Down syndrome"*: Ricki Lewis, "Disappearing Down Syndrome, Genetic Counseling, and Textbook Coverage," PLOS.org, *DNA Science Blog*, August 13, 2015.

98 *CCS lowers miscarriage rates*: William B. Schoolcraft, Mandy G. Katz-Jaffe, "Comprehensive Chromosome Screening of Trophectoderm with Vitrification Facilitates Elective Single-Embryo Transfer for Infertile Women with Advanced Maternal Age," *Fertility and Sterility* 100, no. 3 (2013): 615–19.

4. Silencing a Gene

Much of the material in this chapter is derived from interviews with Jeanne Lawrence, Maureen Gallagher, Brian Skotko, Carolyn Hintlian, and Melanie McLaughlin.

102 *managed to silence the extra copy of chromosome 21*: Jun Jiang et al., "Translating Dosage Compensation to Trisomy 21," *Nature* 500 (2013): 296–300.

102 *numerous papers showing that relatives*: Brian G. Skotko et al., "Family Perspectives About Down Syndrome," *American Journal of Medical Genetics Part A* 170, no. 4 (2016): 930–41; Brian G. Skotko et al., "Having a Brother or Sister with Down Syndrome: Perspectives from Siblings," *American Journal of Medical Genetics Part A* 155, no. 10 (2011): 2348–59; and Brian G. Skotko et al., "Having a Son or Daughter with Down Syndrome: Perspectives from Mothers and Fathers," *American Journal of Medical Genetics Part A* 155, no. 10 (2011): 2335–47.

103 *previous-generation tests that measured levels*: "Screening for Fetal Aneuploidy," Practice Bulletin 163, American College of Obstetricians and Gynecologists Committee on Practice Bulletins—Obstetrics, Committee on Genetics, and the Society for Maternal-Fetal Medicine, May 2016.

104 *Down syndrome was one of several suggestions*: O. Conor Ward, "John Langdon Down: The Man and the Message," *Down Syndrome Research and Practice* 6, no. 1 (1999): 19–24.

104 *no more genetically likely to have an extra chromosome*: Ibid.

104 *figured out what causes the condition*: Eric Pace, "Dr. Jerome Lejeune Dies at 67; Found Cause of Down Syndrome," *The New York Times*, April 12, 1994.

104 *it was actually she, and not Lejeune*: Beryl Lieff Benderly, "A Scientist Saint?," *Science*, February 24, 2015.

106 *She wasn't old enough*: In 2007, the American College of Obstetricians and Gynecologists said that any woman, regardless of age, should have the right to any screening or diagnostic test she desires.

108 *enrolled James in the study for a drug*: "Down Syndrome Program: Clinical Trials," Massachusetts General Hospital, http://www.massgeneral.org/children /research/down-syndrome/ELND005-trial.aspx.

110 *Skotko was also involved in testing a drug*: "Roche: Clinical Trial for Children with Down Syndrome, Ages 6–11, Looking at the Effects of a New Medication on Learning, Memory, and Language," Massachusetts General Hospital, http://www.massgeneral.org/children/research/down-syndrome/down -syndrome-roche-clinical-trial-for-children-with-down-syndrome-ages-6-11- looking-at-effects-of-a-new-medication-on-learning-memory-and -language-q-and-a.aspx.

110 *pursuing a different drug*: Angela Townsend, "CWRU Down Syndrome Researcher Has Personal Stake in Potential Breakthrough Studies," *The Plain Dealer*, April 23, 2015.

110 *drug is being studied in Australia*: "Compose: Cognition and Memory in People with Down Syndrome," http://compose21.com/study.htm.

110 *Prozac administered to women pregnant*: Bonnie Rochman, "Parents Turn to Prozac to Treat Down Syndrome," *MIT Technology Review*, January 12, 2016.

112 *People with Down syndrome develop Alzheimer's*: "Alzheimer's Disease in People with Down Syndrome," National Institute on Aging, https://www.nia.nih .gov/alzheimers/publication/alzheimers-disease-people-down-syndrome.

112 *Roche discontinued the trial*: "Statement on CLEMATIS Trial," F. Hoffmann–La Roche, June 28, 2016, http://www.roche.com/media/store/statements.htm.

113 *McLaughlin's struggle to reconcile her feelings*: Bonnie Rochman, "Early Decision: Will New Advances in Prenatal Testing Shrink the Ranks of Babies with Down Syndrome?," *Time* magazine, February 27, 2012.

118 *some of the only surveys of how people with Down syndrome*: Brian G. Skotko et al., "Having a Brother or Sister with Down Syndrome: Perspectives from Siblings," *American Journal of Medical Genetics Part A* 155, no. 10 (2011): 2348–59; and Brian G. Skotko et al., "Having a Son or Daughter with Down Syndrome: Perspectives from Mothers and Fathers," *American Journal of Medical Genetics Part A* 155, no. 10 (2011): 2335–47.

119 *"We will beat this disease"*: " '21 Thoughts' by Dr. Jerome Lejeune," Jerome Lejeune Foundation USA, http://lejeuneusa.org/advocacy/21-thoughts-dr -jérôme-lejeune#.V1zw51duDFK.

119 *she is working with cute furry mice*: Bonnie Rochman, "A Change of Mind," *MIT Technology Review*, December 16, 2015.

121 *a frog named Bea*: Brendan Maher, "Personal Genomics: His Daughter's DNA," *Nature* 449 (2007): 773–76.

121 *christened Ts1Cje (after Charles J. Epstein)*: Margalit Fox, "Charles Epstein, Leading Medical Geneticist Injured by Unabomber, Dies at 77," *The New York Times*, February 23, 2011.

5. What Do Parents Want to Know?

Much of the material in this chapter is derived from interviews with Maya Hewitt, who requested that her name and those of her family members be changed, and with Nancy Spinner and Ian Krantz, and Meredith Hardy.

127 *A two-vessel cord*: "Umbilical Cord Conditions," March of Dimes, http://www.marchofdimes.org/complications/umbilical-cord-conditions.aspx.

130 TERT *mutations, or changes in the gene*: "Dyskeratosis Congenita: Genetic Changes," Genetics Home Reference, U.S. National Library of Medicine, https://ghr.nlm.nih.gov/condition/dyskeratosis-congenita.

131 *exome sequencing, which scans just the protein-coding portions*: The exome is but a fraction of the genome—a person's genetic code—but it contains what's assumed to be the most relevant genetic information. Because the cost is lower and there's less data to interpret, whole exome sequencing is often more likely to be performed than whole genome sequencing.

133 *genetic error on chromosome 4*: "Genes and Disease: Huntington Disease," National Center for Biotechnology Information, U.S. National Library of Medicine, http://www.ncbi.nlm.nih.gov/books/NBK22226/.

134 *one in sixty-eight children*: "Autism Spectrum Disorder," Data and Statistics, Centers for Disease Control and Prevention, http://www.cdc.gov/ncbddd/autism/data.html.

134 *sequence the genomes of 10,000 people*: "Autism Speaks and Google Harness 'Cloud' for Genomic Breakthroughs," *MSSNG*, Autism Speaks, June 10, 2014, http://www.autismspeaks.org/science/science-news/autism-speaks-google-harness-'cloud'-genomic-breakthroughs.

134 *studies have observed a considerable probability*: "Autism Spectrum Disorder Fact Sheet: What Role Do Genes Play?," National Institute of Neurological Disorders and Stroke, http://www.ninds.nih.gov/disorders/autism/detail_autism.htm#3082_6.

134 *Autism tends to run in non-twinned siblings too*: Sally Ozonoff et al., "Recurrence Risk for Autism Spectrum Disorders: A Baby Siblings Research Consortium Study," *Pediatrics* 128, no. 3 (2011): 488–95.

134 *identifying at least some types of autism risk in utero*: Raphael Bernier et al., "Disruptive *CHD8* Mutations Define a Subtype of Autism Early in Development," *Cell* 158, no. 2 (2014): 263–76.

134 *greater risk of having a child with autism*: Sven Sandin et al., "Autism Risk Associated with Parental Age and with Increasing Difference in Age Between the Parents," *Molecular Psychiatry* 21, no. 5 (2015): 693–700.

135 *scrutinize Adam Lanza's DNA*: "No Easy Answer," *Nature* 493 (2013): 133.

135 *genes aren't typically the last word*: "What Is Personal Genetics?," Personal Genetics Education Project, Department of Genetics, Harvard Medical School, http://www.pged.org/personal-genetics-101/what-is-personal-genetics/.

137 *recommended genome sequencing to their patients*: Sharon Begley, "Consumers Aren't Wild About Genetic Testing—Nor Are Doctors," *STAT,* February 12, 2016.

139 Time *magazine cover story*: Bonnie Rochman, "The DNA Dilemma: A Test That Could Change Your Life," *Time,* December 24, 2012.

140 *One of the couple's most vexing cases*: Ibid.

141 *research that Krantz and Spinner are leading*: The CHOP PediSeq Project, Children's Hospital of Philadelphia, http://pediseq.research.chop.edu.

141 *A multiyear grant*: "NHGRI Broadens Sequencing Program Focus on Inherited Diseases, Medical Applications," National Institutes of Health, December 6, 2011, http://www.nih.gov/news/health/dec2011/nhgri-06.htm.

141 *mythology of Pandora's box*: "Pandora's Box," Myths and Legends, http://myths.e2bn.org/mythsandlegends/textonly562-pandoras-box.html.

143 *a case of "silent" celiac disease*: Maria T. Greene, "Non-Classic Manifestations of Celiac Disease: Silent and Latent Celiac Disease," Ann and Robert H. Lurie Children's Hospital of Chicago, https://www2.luriechildrens.org/ce/online/article.aspx?articleID=248.

144 *we still might have changes in the family of genes*: "Celiac Disease: Genetic Changes," Genetics Home Reference, U.S. National Library of Medicine, https://ghr.nlm.nih.gov/condition/celiac-disease#genes.

145 *the hospital's inaugural Clinical Genetics Think Tank*: Sarah Bowdin et al., "Recommendations for the Integration of Genomics into Clinical Practice," *Genetics in Medicine,* online May 12, 2016.

146 *took more interest in "predictive genetic testing"*: Daniel S. Dodson et al., "Parent and Public Interest in Whole-Genome Sequencing," *Public Health Genomics* 18, no. 3 (2015): 151–59.

148 *yields a diagnosis only about 30 percent*: C. A. Valencia et al., "Clinical Impact and Cost-Effectiveness of Whole Exome Sequencing as a Diagnostic Tool: A Pediatric Center's Experience," *Frontiers in Pediatrics* 3 (2015): 67.

148 *inherited the genetic mutation for Creutzfeldt-Jakob disease*: "Prion Disease," Genetics Home Reference, U.S. National Library of Medicine, http://ghr.nlm.nih.gov/condition/prion-disease#diagnosis.

149 *symptoms including memory loss, fever*: "Managing Symptoms of Creutzfeldt-Jakob Disease," University of California, San Francisco Memory and Aging Center, http://memory.ucsf.edu/cjd/livingwithcjd/managing-symptoms.

6. The Right to an Open Future

Much of the material in this chapter is derived from interviews with Mike Bamshad, Holly Tabor, Robert Green, Lainie Friedman Ross, Robert Klitzman, Laurie Hunter, Marcia Van Riper, Debra Mathews, Jonathan Berg, and Joanna Fanos.

153 *first to sequence the DNA of an entire family*: Sarah B. Ng et al., "Exome Sequencing Identifies the Cause of a Mendelian Disorder," *Nature Genetics* 42,

no. 1 (2010): 30–35; and Bonnie Rochman, "What Your Doctor Isn't Telling You About Your DNA," *Time*, October 25, 2012.

154 *My46, a Web-based program*: My46, "Make Your Genome Work for You," University of Washington, https://www.my46.org.

154 *"Eventually, everyone will have their genome"*: Bonnie Rochman, "The DNA Dilemma: A Test That Could Change Your Life," *Time*, December 24, 2012.

157 *published first online in 2009*: Sarah B. Ng et al., "Exome Sequencing Identifies the Cause of a Mendelian Disorder," *Nature Genetics* 42, no. 1 (2010): 30–35.

158 *originally known as the "ACMG 56"*: Robert C. Green et al., "ACMG Recommendations for Reporting of Incidental Findings in Clinical Exome and Genome Sequencing," *Genetics in Medicine* 15, no. 7 (2013): 565–74.

158 *"If you fall off your bike and get an X-ray"*: Bonnie Rochman, "What Your Doctor Isn't Telling You About Your DNA," Time.com, October 25, 2012.

159 *authored an eagerly awaited policy*: Lainie F. Ross et al., "Ethical and Policy Issues in Genetic Testing and Screening of Children," *Pediatrics* 131, no. 3 (2013): 620–22; Lainie Friedman Ross et al., "Technical Report: Ethical and Policy Issues in Genetic Testing and Screening of Children," *Genetics in Medicine* 15, no. 3 (2013): 234–45; and Bonnie Rochman, "New Guidelines for Genetic Testing in Children," Time.com, February 21, 2013.

160 *now refers to "incidental findings" as "secondary findings"*: ACMG Board of Directors, "ACMG Policy Statement: Updated Recommendations Regarding Analysis and Reporting of Secondary Findings in Clinical Genome-Scale Sequencing," *Genetics in Medicine* 17, no. 1 (2014): 68–69; and "Anticipate and Communicate: Ethical Management of Incidental and Secondary Findings in the Clinical, Research, and Direct-to-Consumer Contexts," Presidential Commission for the Study of Bioethical Issues, December 2013, http://bioethics.gov/node/3183.

160 *debate has continued to percolate*: Wylie Burke et al., "Recommendations for Returning Genomic Incidental Findings? We Need to Talk!," *Genetics in Medicine* 15, no. 11 (2013): 854–59.

160 *the ACMG revised its guidelines*: ACMG Board of Directors, "ACMG Policy Statement: Updated Recommendations Regarding Analysis and Reporting of Secondary Findings in Clinical Genome-Scale Sequencing," *Genetics in Medicine* 17, no. 1 (2014): 68–69.

161 *resurrected a storied legal case*: "Vitaly Tarasoff et al., Plaintiffs and Appellants, v. The Regents of the University of California et al., Defendants and Respondents," Supreme Court of California, July 1, 1976, https://scholar.google.com/scholar_case?case=263231934673470561.

162 *therapists have an obligation to warn*: "Tarasoff v. Regents of University of California," *Casebriefs*, http://www.casebriefs.com/blog/law/torts/torts-keyed-to-dobbs/the-duty-to-protect-from-third-persons/tarasoff-v-regents-of-university-of-california/2/.

162 *the law has wrestled with this issue*: Kristin E. Schleiter, "A Physician's Duty to Warn Third Parties of Hereditary Risk," *American Medical Association Journal of Ethics, Virtual Mentor* 11, no. 9 (2009): 697–700.

162 *Hunter was stunned by the news*: Bonnie Rochman, "The DNA Dilemma: A Test That Could Change Your Life," *Time*, December 24, 2012.

163 *first-person essay that accompanied a series*: Bonnie Rochman, "The Trouble with My Daughter's DNA," Time.com, October 24, 2012.

165 *research study that would examine families' experience*: Marcia Van Riper, "Family Experience of Genetic Testing: Ethical Dimensions," cited in Van Riper, "Genetic Testing and the Family," *Journal of Midwifery and Women's Health* 50, no. 3 (2005): 227–33.

165 *In a normally functioning gene*: "Learning About Huntington's Disease: What Do We Know About Heredity and Huntington's Disease?," National Human Genome Research Institute, https://www.genome.gov/10001215/.

165 *In 1983, the gene became the first*: Gillian P. Bates, "History of Genetic Disease: The Molecular Genetics of Huntington Disease—a History," *Nature Reviews Genetics* 6, no. 10 (2005): 766–73.

166 *the disease unspools*: "Huntington Disease," Genetics Home Reference, U.S. National Library of Medicine, https://ghr.nlm.nih.gov/condition/huntington -disease.

166 *Woody Guthrie, who inherited the disease*: "Woodrow Wilson 'Woody' Guthrie and Huntington Disease," *Huntington's News*, The Quarterly Newsletter of the Huntington's Disease Associations of New Zealand, June 2003, http:// www.huntingtons.org.nz/newsletters/Guthrie0603.htm.

166 *he wrote a poem, "No Help Known"*: Francis Collins, "Making This a Land for You and Me," NIH Director's Blog, National Institutes of Health, February 28, 2013; and Jorge Arévalo et al., "Tracing Woody Guthrie and Huntington's Disease," *Seminars in Neurology* 21, no. 2 (2001): 209–23.

167 *three-quarters of at-risk people*: G. Evers-Kiebooms et al., "Attitudes Towards Predictive Testing in Huntington's Disease: A Recent Survey in Belgium," *Journal of Medical Genetics* 24, no. 5 (1987): 275–79; Seymour Kessler et al., "Attitudes of Persons at Risk for Huntington Disease Toward Predictive Testing," *American Journal of Medical Genetics* 26, no. 2 (1987): 259–70; B. Teltscher, "Objective Knowledge About Huntington's Disease and Attitudes Towards Predictive Tests of Persons at Risk," *Journal of Medical Genetics* 18, no. 1 (1981): 31–39.

167 *more than forty CAGs in a row*: "Learning About Huntington's Disease: What Do We Know About Heredity and Huntington's Disease?," National Human Genome Research Institute, https://www.genome.gov/10001215/.

167 *initial testing for Huntington's began in 1986*: Jason Brandt et al., "Presymptomatic DNA Testing for Huntington's Disease," *The Journal of Neuropsychiatry and Clinical Neurosciences* 1, no. 2 (1989): 195–97; Ann-Marie Codori

et al., "Psychological Costs and Benefits of Predictive Testing for Huntington's Disease," *American Journal of Medical Genetics* 54, no. 3 (1994): 174–84.

169 *she was five years old*: Judith Rosenbaum, "But Mommy, Am I Going to Get
Breast Cancer Too When I Grow Up?," *Tablet*, October 29, 2013.

170 *"In the case of predictive testing for childhood-onset conditions"*: Lainie Friedman
Ross et al., "Technical Report: Ethical and Policy Issues in Genetic Testing
and Screening of Children," *Genetics in Medicine* 15, no. 3 (2013): 234–45.

172 *concept of the right to an open future*: Joel Feinberg, *Freedom and Fulfillment: Philosophical Essays* (Princeton, NJ: Princeton University Press, 1994), 77.

173 *"vulnerable child syndrome"*: Morris Green et al., "Reactions to the Threatened
Loss of a Child: A Vulnerable Child Syndrome," *Pediatrics* 34, no. 1 (1964):
58–66.

174 *Jack Shonkoff wrote a commentary*: Jack P. Shonkoff, " 'Reactions to the Threatened Loss of a Child: A Vulnerable Child Syndrome,' by Morris Green, M.D.,
and Albert A. Solnit, M.D., *Pediatrics* 34 . . . 1964," *Pediatrics* 102, suppl. 1
(1998): 239–41.

175 *" 'patients in waiting' "*: Lainie Friedman Ross et al., "Technical Report: Ethical
and Policy Issues in Genetic Testing and Screening of Children," *Genetics in
Medicine* 15, no. 3 (2013): 234–45.

176 *concerns over vulnerable child syndrome may be overblown*: Bonnie Rochman,
"New Guidelines for Genetic Testing in Children," Time.com, February 21,
2013.

176 *childhood diagnosis may cause parents to ramp up the time*: Joanna H. Fanos, *Sibling
Loss* (Mahwah, NJ: Lawrence Erlbaum Associates, Inc., 1996), chapter 2:
"The Family Setting."

7. How to Hunt a Zebra

Much of the material in this chapter is derived from interviews with Vandana
Shashi, David Goldstein, Mike Bamshad, Misha Angrist, Sarah Foye, Alan Beggs,
Kristen Greene, and Kelly Schoch.

180 *showing that sequencing could be successfully employed*: Anna Need et al., "Clinical
Application of Exome Sequencing in Undiagnosed Genetic Conditions,"
Journal of Medical Genetics 49, no. 6 (2012): 353–61.

181 *a rare disease is defined as a condition*: "What Is a Rare Disease?," Genetic and
Rare Diseases Information Center, National Center for Advancing Translational Sciences, National Institutes of Health, https://rarediseases.info.nih
.gov/about-gard/pages/31/frequently-asked-questions.

181 *30 million people living with rare diseases*: "Rare Disease: Facts and Statistics,"
Global Genes: Allies in Rare Disease, https://globalgenes.org/rare-diseases
-facts-statistics/.

182 *Adam's diagnosis took more than a decade*: Bonnie Rochman, "Researchers Solve the Mystery of Child's Illness," Time.com, November 8, 2012.

183 *a gene-sequencing competition that diagnosed Adam*: Bonnie Rochman, "What's Making Adam Sick? A Contest to Sequence Three Kids' Genomes," Time .com, November 6, 2012, and Rochman, "Researchers Solve the Mystery of Child's Illness," Time.com, November 8, 2012.

187 *progress researchers have made in connecting the dots*: Jessica X. Chong et al., "The Genetic Basis of Mendelian Phenotypes: Discoveries, Challenges, and Opportunities," *The American Journal of Human Genetics* 97, no. 2 (2015): 199–215.

188 *The Centers for Mendelian Genomics . . . the Centers for Common Disease Genomics*: "NIH Genome Sequencing Program Targets the Genomic Bases of Common, Rare Disease," National Human Genome Research Institute, January 14, 2016, https://www.genome.gov/27563453/.

189 *MyGene2, a website where families of kids with rare disorders*: JoNel Aleccia, "Needle in the Genetic Haystack: How a New UW Website Is Helping Families, Scientists," *The Seattle Times*, May 7, 2016.

190 *The most baffling diagnostic challenges*: "Undiagnosed Diseases Network," National Human Genome Research Institute, https://www.genome.gov/27550959/.

190 *in 2014 the NIH expanded its program*: Bridget M. Kuehn, "NIH's Undiagnosed Diseases Program Expands," *The Journal of the American Medical Association* 312, no. 6 (2014): 587.

190 *apply to the network on behalf of a patient*: "Undiagnosed Diseases Network," Genetic and Rare Diseases Information Center, https://rarediseases.info.nih .gov/research/pages/27/undiagnosed-diseases-program.

191 *cracked just 25 percent to 50 percent of its cases*: Bridget M. Kuehn, "NIH's Undiagnosed Diseases Program Expands," *The Journal of the American Medical Association* 312, no. 6 (2014): 587.

191 *initial research, published in 2012*: Anna Need et al., "Clinical Application of Exome Sequencing in Undiagnosed Genetic Conditions," *Journal of Medical Genetics* 49, no. 6 (2012): 353–61.

193 *price to sequence a human genome has dropped*: "DNA Sequencing Costs: Data from the NHGRI Genome Sequencing Program (GSP)," National Human Genome Research Institute, https://www.genome.gov/sequencingcostsdata/; and "The Cost of Sequencing a Human Genome," National Human Genome Research Institute, https://www.genome.gov/27565109/.

196 *Brown–Vialetto–Van Laere syndrome was discovered in 1894*: "Brown–Vialetto–Van Laere Syndrome 1; BVVLS1," OMIM (Online Mendelian Inheritance in Man), http://www.omim.org/entry/211530.

197 *results are first confirmed by an outside clinical lab*: "Clinical Laboratory Improvement Amendments (CLIA)," CLIA Law and Regulations, Centers for Disease Control and Prevention, http://wwwn.cdc.gov/clia/Regulatory/default.aspx.

198 *Two of the research papers that Shashi had read*: A. R. Foley et al., "Treatable Childhood Neuronopathy Caused by Mutations in Riboflavin Transporter RFVT2," *Brain* 137, part 1 (2014): 44–56; and Annet M. Bosch et al., "The Brown–Vialetto–Van Laere and Fazio Londe Syndrome Revisited: Natural History, Genetics, Treatment and Future Perspectives," *Orphanet Journal of Rare Diseases* 7, no. 1 (2012): 83.

8. The Genie in the Bottle

Much of the material in this chapter is derived from interviews with Jennifer Garcia, Michelle Huckaby Lewis, Robert Green, Jennifer Puck, Alan Beggs, Jim Evans, Susan Waisbren, Susan Wolf, Aaron Goldenberg, Jennifer Malone Hoskovec, Ingrid Holm, John Niederhuber, Benjamin Solomon, Emma Warin, Katrine Bosley, Sharon Anderson, Michael Glassner, and Marybeth and David Levy.

204 *babies who receive transplants in the first three and a half months*: "Two Researchers Known for Identifying and Treating 'Bubble Boy' Disease Honored by March of Dimes," Immune Deficiency Foundation, Blog, March 27, 2014. http://primaryimmune.org/two-researchers-known-for-identifying-and -treating-bubble-boy-disease-honored-by-march-of-dimes/.

204 *born just one month after SCID had been added*: "Newborn Screening for Severe Combined Immunodeficiency Disorder," Secretary's Advisory Committee on Heritable Disorders in Newborns and Children, http://www.hrsa.gov /advisorycommittees/mchbadvisory/heritabledisorders/recommendations /correspondence/combinedimmunodeficiency.pdf.

204 *two years would pass before Texas would begin*: "Texas Will Screen Newborns for SCID Beginning December 1," IDF SCID Newborn Screening, The Immune Deficiency Foundation's Blog for Severe Combined Immunodeficiency Newborn Screening, http://idfscidnewbornscreening.org/texas-will -screen-newborns-for-scid-beginning-december-1/.

204 *Garcia has become an activist*: "IDF Volunteer Jennifer Garcia Receives Award from March of Dimes," IDF SCID Newborn Screening, The Immune Deficiency Foundation's Blog.

205 *"this little baby changed things"*: "Jennifer Garcia—SCID Newborn Screening in Texas," YouTube, November 1, 2012, https://www.youtube.com/watch?v =2jfG4enf0-A.

205 *from the first moments of life*: "NIH Program Explores the Use of Genomic Sequencing in Newborn Healthcare," National Human Genome Research Institute and Eunice Kennedy Shriver National Institute of Child Health and Human Development, *NIH News*, September 4, 2013, https://www.genome .gov/27554919/2013-news-release-nih-program-explores-the-use-of -genomic-sequencing-in-newborn-healthcare/.

205 *to map out the entirety of babies' genetic code*: One of the hospitals, Children's

Mercy Kansas City, in Missouri, is focused only on using sequencing to aid in rapid diagnosis of very sick newborns.

205 *child's genome is still incomprehensible*: Michelle Huckaby Lewis, "Newborn Screening Controversy: Past, Present, and Future," *JAMA Pediatrics* 168, no. 3 (2014): 199–200.

206 *BabySeq Project, the newborn screening study*: "Genomic Sequencing for Child-hood Risk and Newborn Illness," ClinicalTrials.gov, https://clinicaltrials .gov/ct2/show/NCT02422511.

206–207 *they went into greater detail*: Susan E. Waisbren et al., "Parents are inter-ested in newborn genomic testing during the early postpartum period," *Ge-netics in Medicine* 17, no. 6 (2015): 501–504.

208 *diagnosed the cause of Adam Foye's muscle weakness*: Bonnie Rochman, "Re-searchers Solve the Mystery of Child's Illness," Time.com, November 8, 2012.

209 *"the idea of genetic exceptionalism"*: James P. Evans and Wylie Burke, "Genetic Exceptionalism: Too Much of a Good Thing?," *Genetics in Medicine* 10, no. 7 (2008): 500–501.

210 *open to getting results by phone*: L. Baumanis et al., "Telephoned BRCA1/2 Ge-netic Test Results: Prevalence, Practice, and Patient Satisfaction," *Journal of Genetic Counseling* 18, no. 5 (2009): 447–63.

211 *a friend of mine who survived breast cancer*: Bonnie Rochman, "What Your Doctor Isn't Telling You About Your DNA," Time.com, October 25, 2012.

212 *how test results affect parents*: Susan E. Waisbren, "Effect of Expanded New-born Screening for Biochemical Genetic Disorders on Child Outcomes and Parental Stress," *JAMA* 290, no. 19 (2003): 2564–72.

212 *spearheaded another study*: S. E. Waisbren et al., "Psychosocial Factors Influenc-ing Parental Interest in Genomic Sequencing of Newborns," *Pediatrics* 137, suppl. 1 (2016): S30–S35.

213 *half the states allow teens to independently make decisions*: "An Overview of Minors' Consent Law," Guttmacher Institute, June 1, 2016, https://www .guttmacher.org/state-policy/explore/overview-minors-consent-law.

214 *Genetic Information Nondiscrimination Act (GINA)*: "The Genetic Information Nondiscrimination Act of 2008," U.S. Equal Employment Opportunity Commission, https://www.eeoc.gov/laws/statutes/gina.cfm.

214 *"There's nothing more preexisting than genes"*: Bonnie Rochman, "Why Cheaper Genetic Testing Could Cost Us a Fortune," Time.com, October 26, 2012.

214 *The law suffered a setback*: "Genetic Information Nondiscrimination Act: A Rule by the Equal Employment Opportunity Commission on 05/17/2016," Federal Register: The Daily Journal of the United States Government, https:// www.federalregister.gov/articles/2016/05/17/2016-11557/genetic-information -nondiscrimination-act.

214 *bipartisan Genetic Research Privacy Protection Act*: "Senators Warren and Enzi Introduce Bipartisan Bill to Strengthen Genetic Privacy Protections for

Research Participants," Elizabeth Warren, U.S. Senator for Massachusetts, April 5, 2016, http://www.warren.senate.gov/?p=press_release&id=1096.

215 *cost of newborn screening*: Bradford L. Therrell et al., "Current Status of Newborn Screening Worldwide: 2015," *Seminars in Perinatology* 39, no. 3 (2015): 171–87.

216 *"overwhelming economic health expenditures"*: Jacques S. Beckmann, "Can We Afford to Sequence Every Newborn Baby's Genome?," *Human Mutation* 36, no. 3 (2015): 283–86.

219 *Emma Warin enrolled in the study*: Bonnie Rochman, "Will My Son Develop Cancer?," Time.com, October 22, 2012.

220 *sequencing is not at the point that it can substitute*: Dale L. Bodian et al., "Utility of Whole-Genome Sequencing for Detection of Newborn Screening Disorders in a Population Cohort of 1,696 Neonates," *Genetics in Medicine* 18, no. 3 (2016): 221–30.

222 *try to edit cells in HIV-positive patients*: Amy Maxmen, "Easy DNA Editing Will Remake the World. Buckle Up," *Wired*, August 2015.

222 *try to make nonviable embryos resistant to HIV*: Ewen Callaway, "Second Chinese Team Reports Gene Editing in Human Embryos," *Nature News*, April 8, 2016.

223 *progressive muscle weakness confines kids to wheelchairs*: "Duchenne Muscular Dystrophy (DMD)," The Muscular Dystrophy Association, https://www.mda.org/disease/duchenne-muscular-dystrophy.

224 *Changes to somatic cells aren't passed on*: "Are Chromosomal Disorders Inherited?," Genetics Home Reference, U.S. National Library of Medicine, https://ghr.nlm.nih.gov/primer/inheritance/chromosomalinheritance.

224 *therapy involves replacing her defective mitochondria*: "Mitochondrial Replacement Techniques: Ethical, Social, and Policy Considerations," The National Academies Press, The National Academies of Sciences, Engineering, and Medicine, Washington, DC, February 3, 2016, http://www.nationalacademies.org/hmd/Reports/2016/Mitochondrial-Replacement-Techniques.aspx.

224 *nearly twenty "three-parent" babies were born*: Kim Tingley, "The Brave New World of Three-Parent I.V.F.," *The New York Times*, June 27, 2014; and Garry Hamilton, "The Hidden Risks for 'Three-Person' Babies," *Nature* 525 (2015): 444–46.

224 *took the nucleus from the mother's egg*: Jessica Hamzelou, "World's First Baby Born with New Three-Parent Technique," *New Scientist*, September 27, 2016.

225 *the United Kingdom passed legislation*: Gretchen Vogel and Erik Stokstad, "U.K. Parliament Approves Controversial Three-Parent Mitochondrial Gene Therapy," *Science*, February 3, 2015.

225 *an expert panel assembled*: "Mitochondrial Replacement Techniques: Ethical, Social, and Policy Considerations," The National Academies Press, The National Academies of Sciences, Engineering, and Medicine, Washington,

DC, February 3, 2016, http://www.nationalacademies.org/hmd/Reports /2016/Mitochondrial-Replacement-Techniques.aspx.

225 *federal law . . . precludes the FDA from research*: Glenn I. Cohen, et al., "Preventing Mitochondrial DNA Diseases," *JAMA* 316, no. 3 (2016): 273.

225 *Speaking in one voice in 2015*: Nicholas Wade, "Scientists Seek Moratorium on Edits to Human Genome That Could Be Inherited," *The New York Times*, December 3, 2015.

226 *apply CRISPR to embryos*: Nicholas Wade, "British Researcher Gets Permission to Edit Genes of Human Embryos," *The New York Times*, February 1, 2016.

226 *why not construct artificial ones?*: Ike Swetlitz, "Top Scientists Hold Closed Meeting to Discuss Building a Human Genome from Scratch," *STAT*, May 13, 2016; and Andrew Pollack, "Scientists Talk Privately About Creating a Synthetic Human Genome," *The New York Times*, May 13, 2016.

226 *Human Genome Project–Write*: Andrew Pollack, "Scientists Announce HGP-Write, Project to Synthesize the Human Genome," *The New York Times*, June 2, 2016.

226 *"sequence and then synthesize Einstein's genome"*: Drew Endy and Laurie Zoloth, "Should We Synthesise a Human Genome?," *Cosmos*, May 12, 2016.

226 *how exactly to impose structure and regulations*: Debra J. H. Mathews et al., "CRISPR: A Path Through the Thicket," *Nature* 527 (2015): 159–61.

227 *poll of 1,000 U.S. adults*: *STAT*-Harvard Poll: Sharon Begley, "Americans Say No to 'Designer Babies,'" *STAT*, February 11, 2016.

227 *gene editing on an annual worldwide threat assessment*: Antonio Regalado, "Top U.S. Intelligence Official Calls Gene Editing a WMD Threat," *MIT Technology Review*, February 9, 2016.

228 *lauded as the first baby in the world*: Ian Sample, "IVF Baby Born Using Revolutionary Genetic-Screening Process," *The Guardian*, July 7, 2013.

Acknowledgments

On my desk, I keep a cartoon captioned "How Reporters Start Their Day at Work . . ." A gaggle of journalists stands beneath a sign that reads *"Today I Am An Expert In:"* One of the reporters is covering his eyes and launching a dart at a bulletin board, which is festooned with potential topics du jour: "Politics. Economy. Car Repair. Health Care. Oil. Vinegar. Stocks. Bondage." There are many other spellbinding selections (septic tanks, anyone?), but the point is clear: journalists can't be experts in everything. So how in the world can we write about such a vast menu of subjects?

We surround ourselves with smart people. "Grateful" barely scrapes the surface of how I feel about the vast number of smart people who helped bring this book to fruition. Meredith Hardy described the army of doctors that helps keep her sons in working order as "celebrities," and I concur. The corps of ethicists, physicians, genetic counselors, and scientists who patiently explained their work to me in granular detail are my own personal rock stars. They are a who's who of influencers in the field of reproductive and pediatric genetics and bioethics. I remain stunned by their insights and appreciative of their time. Though I couldn't incorporate everyone I interviewed, each conversation helped sculpt and inform the scope of this book. If I named everyone who helped me, I would need an additional chapter. But a special shout-out is due Shivani Nazareth, Holly Tabor, Brian Skotko, and especially Mike Bamshad, who read (and, in Mike's case, re-read, with nary a complaint) multiple chapters to ensure technical accuracy. Heather Mefford graciously allowed me to bunk with her at a genetics conference. And Ben Wilfond granted me a long-term loan from his personal library, handing over a stack of books about bioethics as I began this project.

While the many experts who shared their insights into how genetics is reshaping the way we parent are the thought leaders, the mothers and fathers whom I interviewed are the emotional heart of this book. Dozens of families welcomed me into their homes and shared their experiences with genetic testing, their DNA dilemmas. I have played with their children, eaten dinner with them, and asked questions and more questions to which they responded with grace and patience.

When I think of these families, their strength and persistence and determination to do right by their children are top-of-mind. One family, the Belchers of Blue Bell, Pennsylvania, left a particularly lasting impression upon me. Their daughter, Juliet, is confined to a wheelchair. She has a progressive, degenerative genetic disorder and can't walk, talk, read, write, or eat as others do. Because her body has trouble maintaining a stable temperature, Juliet can't go outdoors in the summer and winter. So her parents brought the outdoors inside, painting murals of trees and flowers and glittery bumblebees on her bedroom walls. They tirelessly support researchers in their quest to understand what's gone awry in Juliet's genes, believing that every day brings them closer to a treatment. What others might consider hopeless, Janis and Mike Belcher find hopeful. Janis wears a necklace inscribed with Juliet's name and birthday on a charm; a smaller charm that overlays it is inscribed "Hope." I wrote this book for parents like the Belchers—the parents of a sick child—and for the parents of yet-to-be-conceived children who turn to genetic advances to help them have healthy kids, and for all those parents who fall in the vast middle place, in between sick and well. We all want healthy kids, but we don't all get them. For the latter group, there is hope, and never has it been more promising than in the era of genomics.

I am grateful to numerous people at *Time*, starting with John Huey, who first stepped up to mentor me twenty-five years ago and emboldened me to take on this project. Thank you to editors Nancy Gibbs, Cathy Sharick, and Tom Weber, who supported the reporting that inspired this book. Thank you to Sora Song, who never got annoyed when I wrote yet another story about how genetics was impacting parenting, and to Julie Rawe, who skillfully and tirelessly midwifed into existence an online series and a magazine cover story about sequencing children's genomes. The snarky "WTF" buzzer that she gifted me during the course of that adventure has gotten a real workout as I've muddled through the process of writing my first book.

Boundless thanks to Will Lippincott, my peerless agent, who kept me sane throughout this process, seamlessly transitioning from the role of literary agent to psychotherapist to ace negotiator and back again. He found this book the perfect home at Scientific American / Farrar, Straus and Giroux, where I was fortunate enough to have not one but two devoted editors direct their talents toward my text. Thank you to Amanda Moon for kicking off the process with an abundance of incisive questions and no shortage of enthusiasm for the topic, then returning from maternity leave to seamlessly resume her role as fearless leader. Her insights were only made sharper by her pregnancy. My husband noted that writing a book is not too different from giving birth—both considerable acts of creation. (Amanda, you got the better deal, gestating for just nine months!) My deep gratitude extends to Alex Star for doing the heavy lifting in the home stretch, nimbly diving into a complex topic, asking just the right questions, and gracefully shaping my too-long text into a coherent, cohesive narrative. Editorial assistant Scott Borchert patiently guided me through FSG's book-building process. Copy editor Annie Gottlieb focused her eagle eye on my manuscript, checking facts and making each sentence sing. Proofreaders Judy Kiviat and Debra Fried and production editor Scott Auerbach took perfectionism to new heights, scrutinizing every single word. And Bobby Wicks devised no shortage of creative ways to spread the word about the finished product.

Big thanks to my husband, Dov Pinker, who encouraged me to write this book and who has offered unconditional support. I am grateful for my parents, Judy and Steve Rochman, who flew across the country more than once to take care of their grandkids while I was on deadline. Thanks to Janine Zacharia for introducing me to Will and offering her invaluable unvarnished feedback and support every step of the way. And a huge debt of gratitude to my children, who threw "Writer Mama" a party and wrote me a song when I reached my first milestone in 2015, submitting the initial 100 pages of this manuscript. They have served as my indefatigable cheerleading squad, even as I've cloistered myself upstairs with the admonition "Do not interrupt me unless there's blood." There never was so much as a drop, and for that too I am grateful.

Index

A Note About the Author

Bonnie Rochman is an award-winning journalist. A former health and parenting columnist for Time.com and staff writer for *Time* magazine, she has written for *The New York Times Magazine, The Wall Street Journal, MIT Technology Review, Scientific American*, and *O, The Oprah Magazine*. She lives in Seattle with her husband and their three children. Follow her on Twitter at @brochman and visit her website at www.bonnierochman.com.